Haunted

Independence, Missouri

By Margie Kay

Published by Un-X Media
PO Box 1166
Independence, MO 64051
www.unxmedia.com

ISBN-13: 978-0-9988558-4-4

Printed in the United States of America
Third Edition
(Previous editions were named Haunted Independence)

Contact Margie Kay at margiekay06@yahoo.com
or write to the above address.
Website: www.margiekay.com

Dedicated to my best friend, Donna Hartman
my friend Jamyi T.,

and my daughter, Maria Christine

And special thanks to my husband, Geno,
for putting up with me and my ghosts.

Table of Contents

Introduction

One of my earliest experiences with ghosts was when at the age of 14 I experienced a house we moved into that turned out to be very haunted. I knew I was psychic long before that, and had seen and communicated with ghosts, but this was the first time I experienced a lot of ghostly activity. The 1850 house is in Springfield, Missouri near the Drury college campus. I felt a very cold chill the first day we walked in the door, and my mother felt it, too. We both thought the house might have several spirits attached to it.

Our family took all day to move in, so by evening we were very tired and went to bed around 9:00 PM. Not long after going to bed, we heard my sister scream. My father, brother, and I rushed to her room at the far end of the hall. We all saw someone lying on Alice's mattress on the floor. I called out—"Alice?" She answered from the sun room that was 10 feet away. A ghost slowly disappeared from the bed, letting the blanket fall down it did so. Alice slept in the sun room after that.

The remainder of the two years my family spent in that house was busy with daily sounds of people stomping up and down the stairs, someone running the elevator, men walking across floors, moving drapes, and opening doors and windows.

I saw multiple three-dimensional faces in the walls of the house, and witnessed a murder scene in my bedroom mirror. I couldn't wait to leave, but at the same time had an overwhelming curiosity about who these people were and why I was seeing them. Living in this house made me realize that there was definitely more to this world than three dimensions.

While I lived in Springfield, Missouri my ninth-grade science teacher introduced ESP to the class. I volunteered to be a test subject and we quickly found out that my scores were off the chart. The teacher did tests outside of school hours for about a month, and then he stopped because my accuracy frightened him. He did help me to investigate my house though, and we found out that it used to be a hospital—hence the elevator and multiple ghosts. Years later, I visited the house again long with my friend, Donna. She and I both saw an old woman dressed in a long black dress come into view and then disappear. Evidently the house is still haunted!

It was at age 14 that I became fully aware that I had a special ability, and decided to research ESP as much as possible. One of the first books I read was by Edgar Casey, the well-known psychic. After reading that book I knew that I was like him and the rest of my life would be anything but boring.

Both of my daughters showed psychic abilities at very young ages. I neither encouraged nor discouraged them to develop their sixth sense—I don't think there would have been much point in trying to influence them anyway as I believe this happens naturally, and is probably genetic. My father was psychic and now my grandchildren are showing advanced awareness, so you can see where my opinion about genetics comes from. Someday I am sure that a scientist will find the "6th Sense" gene.

For 30 years I have done paranormal investigations, which include ghost hunting and other unexplained phenomena. I have worked on a number of burglaries, homicide, and missing persons cases with law enforcement and was able to assist in finding most of the missing persons in these cases with a high degree of accuracy. The dead are the ones who lead me to their bodies and perpetrators.

Talking to the dead is similar to talking to a live person. I believe that the dead are in fact, aliv--, but in spirit bodies rather than physical bodies. I've been fortunate enough to speak with a lot of dead people, who have given me a much information about their realm.

Please also read my book "*Gateway to the Dead: A Ghost Hunter's Field Guide*," which describes my views on what ghosts are, what the different types of entities are and how to handle them and protect yourself spiritually while ghost hunting.

I think sprits are fascinating. They can teach us a lot about other dimensions, help us with our daily lives, and assist us in finding out why they haunt. In this book I relate our investigations of haunted locations. My methods are a little different than most investigators; by using both psychic and scientific means I can usually provide a more thorough explanation for a haunting.

I hope you enjoy reading this book about haunted sites in and near my hometown of Independence, Missouri.

Margie Kay

A Brief History of Independence

Independence, Missouri in 1909, by Frederick J. Bandholtz

Independence, Missouri, located just east of Kansas City and next to the Missouri River, was first a Native American settlement, then a large pioneer settlement. Later it was the launch site of the Santa Fe Trail starting in 1821 as a commercial and military highway, until the arrival of the Santa Fe railroad in 1880. Since the river boats could not travel any further up the Missouri River, Independence was a natural stopping point for travelers going west. The city was the hub for the California Trail from the 1840's to the 1860's, and the Oregon trail between 1841 and 1869. Thousands of wagons, teams of horses and oxen, livestock and supplies passed through Independence.

In the 1840's the towns of Westport, Independence, and Kansas City merged to make a larger city called Centropolis, but later Westport became part of Kansas City and Independence became the largest suburb of Kansas City with over150,000 people living here today.

The large, free natural spring located at the site of the current Frontier Trails Museum was heavily utilized by pioneers heading west. The old wagon trail can still be seen in some spots.

The 160-acre site known as Big Spring, as it was called by the Native Americans who lived in this area, became the Town of Independence on March 29, 1827. The town thrived due to the influx of pioneers headed west. The courthouse was built in 1837, and more permanent brick buildings soon began to replace the wood frame structures soon after.

Two Civil war battles were fought here—one on August 11, 1862, when the Confederate army captured the town, and another in October of 1864 when a bloody battle was fought for two days near what is now Crysler and Lexington streets at the railroad cut. There were also many other skirmishes in the area. At least one civil war hospital was located here (and is still standing). The famous "Order No. 11" that displaced hundreds of families in Jackson County included many that lived in Independence.

There was fierce debate over the slavery issue in Missouri, and much fighting and bloodshed occurred in this location. It is no wonder that there are still many Union and Confederate spirits of soldiers and their families haunting the area.

The Bingham-Waggoner estate was built in 1855 and the Vaile mansion was built in 1881. Both are reportedly haunted, and I talk more about that later in the book.

Independence has been home to Ginger Rogers, Joseph Smith, Hirum Young, President Harry Truman, Bess Truman, Albert Pujols, George Caleb Bingham, and for a while, Frank and Jesse James and Wild Bill Hickok, among others..

The Independence Chamber of Commerce hosts a large festival called Santa-Cali-Gon every year in honor of the men and women who braved the trails.

When I drive through my hometown of Independence, especially at night when it is quiet, I sense many spirits. I am sure that most homeowners are unaware of their ghostly roommates. Most people think that hauntings are an unusual occurrence, but I have found the opposite to be true, and that they are fairly common. Perhaps your house is haunted, too.

The following locations have all been visited by the QUEST Paranormal Investigation Group and are haunted. However, not all places are open to the public. The locations that are open to the public are listed as such. When visiting graveyards, check for closing times and don't visit after dusk or you'll be chased off the property. Don't go anywhere on private property without permission from the owners.

Part I
Haunted Places
Open to the Public

1850 engraving made for a German publication.

Haunted Ricky Road Legends

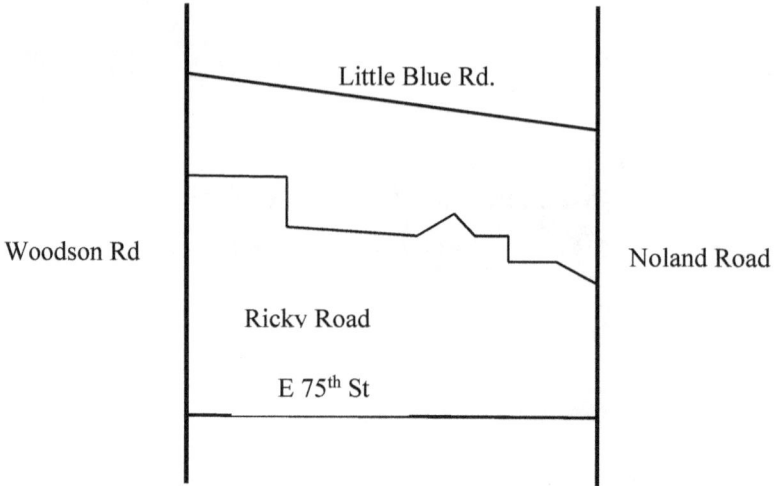

Located between Independence and Raytown, between South Noland Road and Woodson Road, also South of Little Blue Road in Independence, Missouri; Ricky Road is dark, narrow, and has many curves, twists and turns, and is famous for its Urban Legends. It is definitely the scariest road in Independence.

One submitter on the website Forgottunus.com says that "Sleepy Hollow has nothing on this place." The poster says that one legend refers to a girl who was killed here in the 1960's and was beheaded, and that some people have seen her head roll across the road. Stories abound about a teenager who hung himself under the bridge, and a child who was killed by his stepfather, who built concrete steps to the house the next day and buried the child in the steps. The word M-U-R-D-E-R was painted on the road signs in front of the house where this happened.

Some people have reported seeing a woman wearing a white wedding dress and veil, who stands on the side of the road and stretches her arms towards cars. Another person reported seeing a decapitated head with a horrible expression on its face roll out of the woods and under the front of the car, but when he stopped to investigate nothing was there.

There are street lights along most of the road now, which makes it easier to see, but the road is a scary place. Trees are overgrown and almost cover the road overhead in places, and there are driveways that lead to vacant lots, and stone walls that surround overgrowth. Since there have been quite a few auto accidents on this challenging road, it is not surprising that a ghost or two has been spotted along this one and one-half mile stretch of road.

One evening in October, 2003, my daughter and her friend were driving on Ricky road to see if they could locate a house that was reported to be haunted, when they spotted a woman walking on the side of the road. As they approached they both could see that she was glowing and transparent, and had no feet. A few months later, my daughter again found herself on this road and saw a glowing ball of light moving ahead of her.

Another witness drove on this road with a friend after dark in 1980 and as they approached the top of a hill the driver turned the lights out. They could see a fire burning in the distance, and a group of people dancing around the fire and making odd motions. The witnesses got closer, then stopped and got out of the car, but upon doing so noted a "very heavy, eerie feeling and sadness." They got back in the car immediately, turned around and went back the way they came without further investigation. Another witness drove along the road one evening and suddenly had a strange, eerie feeling that he was being watched, when the car died right on a sharp turn and would not start again. He called his parents by cell phone (with poor reception) and as soon as they arrived the car started without a jump. They never could find an explanation for it.

Tales are told about unexplained creatures running along the side of the road, cemeteries appearing and disappearing, and stone walls that appear and disappear.

The QUEST team visited Rickey Road in August of 2010 late at night. A dog chased our car, but it was not a ghost dog. We saw a skunk, an opossum, and a cat that did not move when we drove next to it. We also saw two white foggy mists that could not be explained. On the way back we saw a very large snake in the middle of the road and I pulled the car to the left to avoid it. My assistant took a photo and later identified the snake as a copperhead. We thought it was odd because copperheads are usually seen around water and there is no water near the area. In any case, the snakes are usually found hiding in woodpiles and old abandoned buildings, not in the middle of a road. We did not see any ghosts that night, but definitely got a creepy feeling about the place.

Be careful if you drive on this road because it is difficult to navigate. I'd suggest a run-through during the daytime first to get familiar with the road before going after dark.

The Historic Truman Depot

600 S. Grand Ave., Independence, MO 64050

This historic depot is where Harry Truman was met with 8,500 admirers upon his return after leaving office in 1953. There have been reports of eerie music playing at the depot at night, and no source can be located.

Since our move to Independence in 1987 I noticed a problem with a railroad crossing near my home, not far from the depot. It seems to go off of its own accord, even when there is no train in sight. Sometimes the bell continues ringing for a long time, and I have called the railroad on a number of occasions to complain. However, it may not be the fault of the railroad.

Recently, while approaching the intersection at Scott and 20th Street at approximately 8:00 a.m., I noticed that the red crossing lights were on. I looked down the tracks to the East and saw a train with a small puff of smoke billowing out and its big bright front light on. It appeared to be about a quarter-mile away and moving very slowly. After crossing the track, I looked back but no train was in sight! I realized then that the problems with the crossing probably had nothing to do with the current railroad trains.

This railroad cut and set of tracks dates to the Civil War era, and is the same track that goes by the Lexington house mentioned earlier in this book. If you visit, watch out for a transparent steam locomotive at the old Independence Depot. Several people told me about the ghost train, so I did my own investigations. I've seen this train twice while crossing the bridge on Lexington. The train looks perfectly intact, and is complete with smoke billowing out the top, but you can see through it. The depot is very near the Haunted Lexington street location mentioned in this book.

I wonder if it is the same train traveling by my house at the Scott intersection. It might be worth a trip to the depot to see if you can see the ghost train for yourself.

The Noland Road Lady in Gray

Woodlawn Cemetery, possible resting place of the Lady in Gray

The Gray Lady appears on North Noland Road to South Noland Road, and along 23rd Street. I thought this was just an urban legend until I saw the Noland Road Lady in Gray myself in the summer of 2010. There she was, walking on 23rd Street headed east, just a few blocks from Noland Road on the South side of the street. Dressed in a long grey skirt and blouse, in tatters, the pale-faced woman was looking around as if searching for something. I thought "She *must be the Noland Road Lady in Gray*," when suddenly, as I drove by, she disappeared into thin air. My suspicion was confirmed.

I'd heard the stories for years—how the ghost of a woman walking on and near Noland Road who appeared and disappeared before many people's eyes. And how this lady with short gray hair (or pinned up hair) frequented the area around the large Woodlawn Cemetery, how she usually appeared after dark, but sometimes appeared at dusk as well. Some witnesses told me that she has a very white face and large eyes, which was indeed what I saw as well.

Rachel C. saw the lady one day while driving north on Noland Road in the summer of 2011. She drove past, then Rachel saw the lady again just a few minutes later two miles further north. The woman was walking on the side of the road and was wearing tattered grey clothing and had dark grey hair. She seemed to be looking around for something. "There is no way that a person could outrun a car," said Rachel. "I think this was definitely the famous Lady in Gray." Rachel was quite unnerved by the incident, having never seen a ghost before. "I couldn't believe what I saw," said Rachel, "I 'm a believer now for sure!"

A couple posted to HauntedPlaces.org that they may have seen the Lady in Gray in 2016 when they drove around the cemetery looking for the famous apparition. They drove around the large cemetery several times, watching carefully, when they turned the corner at one point and saw an older lady with a scarf around her face with only her big eyes showing. She was wearing a long skirt and a light on her head, and she looked terrified.

The couple could not figure out how the woman could possibly have walked the distance needed by they time they drove around to that location, so they decided to go up the road a bit and turn around and go back. When the couple got back to the same spot the lady was nowhere in sight, and they couldn't figure out how she could have possibly disappeared in that short length of time.

Haunted Buildings
on the Independence Square

Most are open to the public

The Historic Independence Courthouse

A Salon

The owner of a Salon on the Independence Square called me to investigate a possible haunting in her building. She asked to remain anonymous. She and her staff have noticed sounds coming from areas where they knew no one was located. I knew nothing of the building or its history prior to my arrival. I did a walk-through of the lower level and was immediately drawn to the first room on the left side.

When I entered the room I saw a woman dressed in civil war attire with a long plaid dress, white shirt and white capelet. Her hair was pulled up in the back and she had long white ribbons attached to the bun. The front of her dark blond or light brown hair was parted in the front fell to the sides.

I saw the woman doing paperwork related to some type of business. She feels as if she still owns the building and it is her property. Her husband is nearby but in the background. I could see that their living space was in the upper level. The woman told me her name is Mirabelle and her husband's name was Frank. He worked in the front of the business and she worked in the back and handled the finances. She does not mind the new business being there at all—in fact she enjoys watching the employees work.

Since the ghost is doing no harm, the owner wanted her to stay, so I did not ask her to leave.

The Game Café'

107 W Lexington on the Square

The Game Café' has the spirit of a young boy that had been seen and heard by the owner's young daughter, and another child who was visiting with his father. Both children spoke with the spirit.

I visited the property and psychically viewed a young boy around age 6 or 7 with blond hair who died on the site due to illness. The boy used to carry water and care for livestock when the square was bustling with activity and horses were the transportation method in the mid 1800's. It looked like there may have been a stable nearby because I saw horses tied up to posts and lots of straw. It seems that the child relates to other children and will appear to them more easily than to adults.

The Courthouse Exchange

113 W Lexington on the Square

The Courthouse Exchange is one of the most haunted sites in the city. It is haunted by prankster ghost children in the bar and restaurant area. The manager told me that the ghosts play a "dime game" and leave several dimes, always face-up, on tables and the bar some nights, which the staff finds in the morning. They've also been known to put the T.V. on in the early morning, which is always tuned to a cartoon channel, and move cabinet doors and pictures on the walls. One of the pictures in the women's bathroom is sometimes found upside down or off the wall altogether, even though when the staff leaves the night before they make sure the picture is in the right place.

A ghost girl, who is often mistaken for a living girl, has been spotted by several people who have spoken to me about this over the years. The child plays around some of the tables, and then disappears in front of the patron's astonished eyes. "I saw the ghost of the girl in 2003," says Sadie Miller, of Independence. "She walked over to our table and I thought she was a live child until she skipped away, and I watched as she slowly disappeared." Mrs. Miller says that was the spookiest experience she ever had, but that she will continue to dine at the Exchange because the food is so good.

A little boy has been seen in the back room of the restaurant where private parties are held. He walks through the back wall of the building according to some of the staff members.

I ate dinner at the Courthouse Exchange one evening when I saw an apparition of a girl in an 1800's dress skip by our table. She was transparent, but clearly visible. No one else seemed to see the child.

I can recommend dining at the Courthouse Exchange because the food is very good, and perhaps you'll catch a glimpse of one of the ghosts while you are there.

Note: Access to the restaurant, which is on the lower level, is by stairs from the front, or through the back door for handicapped persons.

Café' Verona
And the Lady in Blue

206 West Lexington on the Square

The Café is sometimes visited by a ghostly Lady wearing a blue dress, who is often seen sitting at a table by the window. Often money goes missing from this particular table, so make sure you hand money to your waitress if you sit there! The employees and chef say that in the kitchen doors often open and close on their own, and they hear unexplained noises in the restaurant late at night. One chef even quit because of the ghostly activity in the kitchen that got out of hand one evening.

Patrons have reported seeing an older lady in a long blue dress who purposely spills alcoholic beverages that are left behind on tables. Perhaps she is a relic of the Temperance Movement. The food is quite good- so its worth a visit even if the ghost doesn't appear.

Corporate Copy Print

111 South Main Street

The owners of Corporate Copy Print have experienced unexplained noises and sounds of people running around in the upper level of the building, and loud stomping on the stairs. While working late one night after moving in doing copies, Susan Waters heard drilling or a chain saw upstairs or outside, which was very loud. She went around the building and checked everywhere, but found no source for the noise. She then went next door to ask the neighboring store owners if they heard it, but they said they didn't hear a thing. On other occasions, when she works late, Susan hears people walking around upstairs, but no one is ever there.

I spoke with Bob White, a paper salesman who provides paper for the Corporate Copy Print. He told me, wide –eyed, that he arrived one day to show Susan some new paper. Bob sat in a chair in the front lobby while waiting for Susan, and heard loud banging or hammering coming from the upper level for several minutes.

At that time the building was being renovated and stairs to the upper level were accessible and open. Bob asked Susan who was making noise upstairs, and she replied that no one was upstairs, and the noise was probably caused by their resident ghost. White didn't believe her, so they both went upstairs to look around, but saw nothing and no one in the unfinished empty space. This unnerved Bob, who still makes sales calls at Corporate Copy Print, but he doesn't linger long.

The Square

The Independence Square from the corner of Main Street and Lexington

Watch for Harry Truman's ghost walking about the square day or night with his signature cane! Harry's first job was at Clinton's Soda Fountain on the square, and the 1827 Log Courthouse nearby was where held court in the 1930's as a judge. His future wife, Bess, worked at the library on the square at 208 West Maple, and is where Harry often visited her. Truman's house is just a few blocks from the square, where he took daily walks after his presidency concluded in 1952 until his death in 1972.

Some residents claim to see the spirit of Wild Bill Hickok walking the streets at dusk in his long frock coat and black hat with a sweeping brim, with two Navy Colts placed in a red sash around his waist. Hickok was not only known as a cool, deadly gunfighter, but also for his negotiating skills. The story goes that Hickok got his nickname from a woman in 1862 on the streets of Independence. He had just stopped an angry mob outside of a Saloon on the Independence Square when a woman shouted, "Good for you, Wild Bill." And the name stuck.

Independence was once a true picture of a Wild West town complete with saloons, brothels, outfitters, general stores, livery stables, the county jail, gunfighters and outlaws. With the rich and intense history of the Square, it is not surprising that there are so many ghostly encounters here!

Wild Bill Hickock
Photo: Public Domain – Wikipedia Commons

The Community of Christ Temple

1001 W. Walnut, Independence, MO 64050
816-833-1000
www.cofchrist.org
Tours and Museum open Monday through Saturday

Once known as the RLDS (Reorganized Church of Latter-Day Saints), this is the headquarters for the church, near where Joseph Smith prophesied in

1832 that a temple would be built in Independence in a specific location. The current location is apparently not exactly the one that Joseph Smith spoke of.

This may be why I saw the ghost of Joseph Smith looking at the Temple one day while I was driving by. He was standing on the steps of a nearby unrelated church and looking in the direction of the temple with an angry look on his face. The apparition was all black from head to toe (as a shadow person would), and was wearing a wide-brimmed hat and frock coat.

Reports from staff members who work at the Temple and the Auditorium are that they hear unexplained voices and footsteps in both buildings, especially when there are no visitors and it is quiet. The source for the noises can never be located. One staff member told me that he has tried to follow mysterious footfalls many times, but has now given up because he feels that it must be a ghost that he'll never find.

In 2006, as I was driving home after work one evening, I saw a very strange cloud formation cover the entire temple, almost to the ground. This dark cloud was perfectly rounded in an oval shape. The cloud moved up very quickly to just cover the top of the spiral as I stopped the car. I took a picture with my camera phone, but was unable to get snap the photo before the cloud moved up. A white beam of light then appeared on the top of the spiral, shooting at a 45-degree angle to the right of my position. The light went out or disappeared after a minute. No source or explanation has been given for this.

I receive many reports of paranormal activity from all over the U.S. and other sites. One thing that perplexes me is that I've received at least six reports from persons who live on streets near the Temple who say they've seen strange creatures including monkey-like creatures crawling on the street, and two separate reports from people who saw alien-like beings staring into their windows. What these things have to do with the Temple, I don't know, but I find it odd that they are in close proximity to the structure.

In 1909, Col. Thomas Swope, his friend, and several family members were murdered in their home on Pleasant Street. Read the book "Deaths on Pleasant Street" by Giles Fowler for more information about this ghastly tale. The brick home, known as the "Swope Mansion," was later purchased by the Church of Christ and razed in order to make room for the Temple and parking lots. Perhaps the ghosts of the murdered Swope family still linger, and are the source of the haunting at the current site. Col. Thomas Swope may be unable to rest thinking of his beautiful brick mansion being razed.

Note: the staff members will likely not speak to visitors about the ghostly activity. Most of the information about the ghostly activity was obtained by persons who wish to remain anonymous.

Interestingly, the Swopes are my ancestors on my Great-grandmother, Minnie Swope's side.

The Music Arts Institute

1010 S Pearl St. Independence, Missouri 64050

Students at the Music Arts Institute may get more for their money than music lessons. When 15-year-old Sara T. took cello lessons there in the mid-1990's she heard the eerie sounds of a second cello playing down the hallway of the building. She thought that she and her instructor were the only two people on the second floor, but when she asked her instructor who else was playing he said it was "just the ghost." The instructor told her that a young girl died there many years ago.

To investigate, I entered the building one evening during open hours and walked the halls. After about 20 minutes, I heard a cello playing very sad music. I walked around, trying to find the person who was playing, but could find nothing. The rooms were empty except for one, where a teacher and student were working, but the child did not have a cello. It sounded like the tones were coming from down the hall, but when I walked that direction, the sounds then came from the opposite direction. I spoke to an instructor later who said that when people try to find the source of the music, they are unable to locate it.

Drive by this building located near 23rd and Pearl Streets, and look up at the third-floor windows. You might catch a glimpse of a young girl with long blonde hair, even after hours. The apparition has been reported by multiple witnesses, and I saw her myself in 1998. I was driving by the building, looked up and saw a young woman standing in the window. It was at 9:30 pm when the building was closed and there shouldn't have been anyone in the building.

In the past few years the City has changed out the windows to a reflective type, likely they are more energy efficient. I have not seen the ghost girl in the windows since, and it may be due to the type of windows that are now on the structure.

Mound Grove Cemetery

1818 N. River Blvd., Independence, Missouri

Some people would say we are nuts for visiting graveyards on Halloween at midnight and they may be right, but what better time to see ghosts that might be lingering? The reason we investigated this location was because of an incident that local resident Rachel D. had in 1994. She was 16 years old and had just had a fight with her boyfriend. Rachel decided to drive to this cemetery to be alone and think. And for whatever reason, she got out of her car and walked through the cemetery at—yes– *midnight* on Halloween night.

R.T. found a nice spot next to a tree and sat down. She was approximately 200 yards from the front gate. No one else was around (who would be?) and it was a very quiet, moonlit night. Rachel closed her eyes and meditated for a few minutes. The thought occurred to her that up to this point she had never actually seen a ghost, although people around her could and she felt somewhat jealous of them. She thought "What better time and place to see a ghost?" and said out loud "If there is such thing as ghosts, show yourself to me right now!"

Right on cue, a white apparition appeared over the hill. Rachel couldn't believe it, but she stood up and stared at it to be sure. It moved towards her. Now she could see that this was the apparition of a woman wearing a long white nightgown. The spirit had no feet and glided through the air. Panicked, Rachel started walking toward the car. The ghost followed. Rachel then started running, and glancing back she saw that the spirit moved faster towards her. Finally reaching the car in the lot across the street, Rachel looked back and watched as the ghost slowly faded away at the gate. It seems she could not go past the gate.

Our investigation: Our team of five including me, Rachel, Mia, Jamyi, and Donna decided to see if we could get the lady ghost to appear again so we could get her on film. We arrived at the cemetery at 11 p.m., once again on Halloween night, and walked in through the front gate. We agreed to stay together at all times for safety. I started by the entrance and filmed graves in the immediate area with my infra-red camera. Mia took the digital camera and took several shots because sometimes we are able to get imaged of orbs or spirits on film that are not visible to the naked eye. This evening we saw a fog and blue lights appear over some of the graves, but the fog and lights are not visible on the film. No apparitions made their appearance at that time.

We moved further into the park, and as we did so it got darker since we no longer had the streetlights illuminating the area. The moon was out so that helped a bit. As our eyes became accustomed to the dark we could see better. After a while, Donna and I decided to stay behind while Mia,

Rachel, and Jamyi walked further on. For some reason, we forgot all about our agreement to stay together. Big mistake.

I looked around at some of the local gravestones and took notes. Donna and I felt the presence of a couple of spirits but nothing materialized. I did some filming, and after what seemed like 45 minutes, I called out for our missing team members. They did not answer. I walked a little way down the road where it was very dark and called again. I heard Mia's voice answer me and saw what I thought was her dark, shadowy figure walking towards me on the small paved road. After a minute, the three women appeared at the top of the hill. I then realized that the black shadow walking towards me and the person who answered me was NOT Mia.

I walked backwards away from the entity, and told the three women to look at the bottom of the hill where they all saw the dark apparition of a man moving towards me. They all saw the dark figure and froze where they stood. I moved quickly back up the other side of the hill from where I came, and when I looked back the figure was gone. The five of us decided we'd had enough and exited the cemetery very quickly.

We have since noted that this cemetery is locked after dusk and have not gone back after dark.

Location where the dark shadow of a man appeared, answered me, and walked towards us.

Location where Rachel saw the floating apparition of a woman

The Truman Home

223 N Main Street, Independence, Missouri
816-254-9929
Open to the Public

This 14-room Victorian home was built in 1886 by Bess Wallace Truman's maternal grandfather, George Porterfield. It was called "The Summer White House," and where Harry Truman lived after his marriage to Bess in 1919 until his death in 1972. Truman's Aunt and Uncle lived across the street at 216 North Delaware, and Harry was a frequent visitor. The interior of the Truman home is exactly as it was in 1972 and kept that way until Bess died in 1982.

Watch out for Harry Truman's apparition to appear, sometimes sitting in his favorite chair in the living room or walking outside the home. The smell of his favorite brandy has been noted by guests on occasion. His ghost has been reported around nearby old downtown Independence, where he took daily walks during the last 15 years of his life after his final term as president ended in 1957.

Harry Truman had his first job at Clinton's Soda Fountain on the Independence Square near this home, and worked as an attorney nearby. It is no wonder that he has been spotted hanging out at his old favorite places.

I took a tour of the home and immediately upon entering the kitchen could sense the presence of a spirit, but it felt like an older woman – possibly Bess or her mother in this place. I was drawn to look in the living room and saw the transparent apparition of President Truman sitting in a chair. He did not look up or acknowledge me as spirits normally do. It seemed like he was simply reliving his past.

Many times, while walking on the Independence Square, I could sense the presence of several spirits walking around, and on two occasions I saw what I thought was a live person who looked like President Truman walking with a cane, but in each case, he gradually disappeared into thin air. There is a local actor who portrays Truman, and who looks just like him. One day this man was at an event and I thought it was Harry's ghost, but when he shook my hand I realized that he was real. I said "Oh I'm glad you're not the ghost," and he said, "You've seen him, too?"

Photo of the living room at the Truman Home taken in the 1930's by Jack E. Boucher.
Photo: Library of Congress.
https://www.nps.gov/hstr/index.htm

The home is now a museum run by the National Parks Service and is open for public tours daily. No tours on Mondays between Labor Day and Memorial Day. Tickets can be purchased at the Truman Visitor Center at 223 North Main Street. Hours 9:300a.m—5:00 p.m. daily except on holidays.

The Sermon Center

201 N Dodgion Street
816-325-7370

The Roger T Sermon Community Center, located at the corner of West Truman Road and North Noland Road has a theatre, weight room, game room, gymnasium, and meeting rooms for special events.

Patrons and staff say they can sometimes hear sounds of a child calling out for its mother, and cold spots in the basement of this building. The staff calls the basement "the catacombs."

Perhaps the ghost is actually coming from the two-room log cabin located on the grounds. The pioneer spring cabin was moved from an Irish community called "Brady Town" to this spot in 1971. There is also a pioneer spring nearby where settlers would get fresh water.

Pioneer Spring Cabin

The National Frontier Trails Museum

318 W. Pacific, Independence, Missouri
816-325-7575
www.ci.independence.mo.us
Open to the public Monday through Sunday

This museum, interpretive center, and research library has much information about the three westward trails, mountain men and trappers, and exhibits about free black men who followed the California trails. Located right in the heart of where the trail activity occurred in the 1900's, the museum also has a life-sized drawn figure of Hiram Young, the famous blacksmith, in the blacksmith shop area.

Also on the grounds is the Chicago and Alton train depot, which was built in 1879. The old 1830's grist flour mill, which was expanded by William Waggoner and George Gates, houses part of the museum. The natural spring used by pioneers is also on the site as well as old wagon swales from the early 1900's.

Among the glass displays, when it is quiet, some patrons have reported seeing the reflection of a woman dressed in pioneer clothing. They usually assume that she is a museum worker dressed in period style, however, when they turn around to see who the person is, there is no one there!

Some of the staff and visitors told me that they have also seen lighted orbs floating through the displays at night, felt unseen and unexplained touches on their shoulder by someone's hand, and unexplained voices at all hours. Most of the activity occurs at night.

It is well worth a visit to the museum, but keep in mind that you may also be greeted by resident ghosts!

The Vaile Mansion

1500 N Liberty

816-325-7430

www.vailemansion.org

Open to the public daily from 9:30 a.m. to 4:30 p.m.

This incredible Second Empire Victorian style mansion in Independence, Missouri was built in 1881 by Colonel Harvey Vaile, who ran the Star Route mail delivery company via overland stage and railroad to Santa Fe, New Mexico from Independence, and the Vaile Pure Water Company.

Vaile was a successful abolitionist lawyer, cattle rancher, investor, and landowner who made a fortune from his various endeavors. The mansion has 31 rooms, nine Italian marble fireplaces, running water, and beautiful painted ceilings, and has been fully restored to its original splendor. Even if you're not interested in ghost hunting, I highly recommend a visit!

The ghost stories about the Vaile evolved around different time periods and events. The first event was when Mr. Vaile was accused of mail fraud and sent to Washington for a trial. Even though he was acquitted, Mrs. Sylvia Vaile, despondent over the accusations, reportedly took an overdose of morphine and died in the home in 1894. Others believe that she may have died from ovarian cancer, but that was a taboo subject at the time and may have been the reason for the suicide tale.

Reportedly, Mr. Vaile could not bear to part with his dead wife, so he had her buried in the front yard with a glass cover over the coffin so he could visit her. During Victorian times, it was not uncommon to have glass windows on coffins. Protests from neighbors later resulted in the removal of the coffin to another undisclosed location nearby.

Colonel Vaile died five years later. He never remarried.

Sylvia Vaile has been spotted looking out of the upper floor windows by passersby and inside the house throughout, as well as walking on the sidewalks in the neighborhood. One neighbor told me he would not walk on the sidewalks near the Vaile Mansion because that is where the ghost walks.

The Vaile Pure Water Co. was operated from the site until just after the turn of the century. The mansion was turned into an inn after Mr. Vaile's death in 1894. Later, the home was used as an asylum and sanitarium, and the heavy metal cages were people were kept are still in the basement. The staff does not like to visit the basement because strange noises and an ominous feeling occur there. The Vaile Mansion also served as a rest home, so one could assume that other people died here.

The mansion was purchased by Roger and Mary DeWitt in the 1960's, and after the death of Mrs. DeWitt in 1983 the mansion was donated to the citizens of Independence.

Spirits and ghostly activity have occurred over the years in several rooms in the house, on the grounds, and in the neighborhood.

The third floor is now off limits to the public (perhaps due to pesky spirits) but during my visit on this floor in 2001 I sensed the presence of at least two spirits. One, a male spirit, was sitting by the back window, but as I stepped into the room, he slowly faded away. A second dark outline of a person stood in a corner for several minutes. My husband and I also smelled cigar smoke, but no one was smoking.

The staff also reports seeing imprints of someone sitting on the freshly made beds in one second floor bedroom when no one has been in the room.

Franklin Cemetery

24001 Blue Mills Rd. Independence, MO

Franklin cemetery located on East Blue Mills Road, east of N. Twyman Rd and North of 24 Hwy. From Winner Road, take US 24 East to N. Twyman Road, turn left or North on Twyman 1.3 miles to CR 8N, also known as East Blue Mills road, and turn right, then go .8 miles. It is very small.

Many people have reported seeing faint lights and orbs of difference sizes and colors, which for some reason only appear during the new and full moon phases. The lights or orbs float from the Northeast side of the grounds and follow the same path every time towards the Western gate.

I stopped in front of this cemetery in March of 2010 at dusk during the new moon to see if the stories were true, and did see a glowing ball of light floating through some of the headstones, but it did not go towards any gates. The orb was approximately 6" in diameter and stayed in the area for approximately 10 minutes. I watched for approximately an hour and a half, but no further activity occurred during that time. It would be well worth a visit.

Elmwood Cemetery

Photo: Brian Hillegas, Creative Commons License

4900 E Truman Road

Kansas City, MO

National Register of Historic Places

Established formally in 1872, this cemetery is the second oldest in Kansas City. The first burial here was in 1840. The cemetery holds the remains of more than 36,000 people, many of them famous, and the cemetery is famously haunted as well. There are a couple of lingering spirits that you should keep an eye out for.

Father Henry David Jardine's spirit still wanders, perhaps looking for his missing grave. Father Jardine also haunts the church where he served – St. Mary's Episcopal Church in Kansas City. He committed suicide in 1886 in St. Louis after allegations of wrongdoing by persons who found out he had served jail time in the past. He was accused of misusing parish funds, drug use, and immoral behavior with young girls. He filed a suite against John Shea, editor of The Kansas City Times, but lost the case. Father Jardine's priesthood was revoked and days before he was to contest the decision he we found dead. It took his family 40 years to get his name cleared so his body could be moved from this location to a Catholic cemetery. Witnesses report seeing his spirit hanging about his old grave in the cemetery as well as his old church – St Mary's at 8th and Walnut.

Another possible lingering spirit is that of George William Shaw. Shaw purchased a burial pot at the cemetery, and then was found two days later hanging from a tree. He was cremated, and then buried here. His slow, trudging footsteps can be heard when it is quiet, according to some witnesses.

Hill Park Cemetery

10499 E 20th Street, Independence, MO

Frank James (d 1915) is buried here with his wife, Ann Ralston James (d 1944) (Ralston street is nearby), two Confederate soldiers, along with Adam Hill and his immediate and extended family members. The cemetery was once a part of land owned by the Hills before becoming a public park.

The small cemetery is surrounded by a stone wall and has an iron gate. It appears to never be closed. This is at the top of Hill Park off of 23rd Street in Independence, but you can get to it more easily from 20th Street and park in a small parking area right next to the graveyard. A misty figure of a person with a white glow walks over the hill nightly. Neighbors say they've seen the bright apparition often. Civil War soldiers have been spotted marching in the park, which used to be an area heavily traveled during the Battle of Westport.

This park is located directly across the street from the haunted Rotary Park, where the ghost of Ann Ralston and others still linger. *Note: See the section on Rotary Park.*

Woodlawn Cemetery

56-acre cemetery with 33,000 graves including graves from the Civil War
and Santa Fe Trail

701 South Noland Road, Independence, Missouri

The earliest burial at this site dates to 1819. This site used to be three
separate cemeteries which were joined into one very large cemetery,
bordered by a stone wall. The cemeteries included land purchased in 1837,
the Hansbrough family cemetery purchased in 1845, and an 1853 purchase
called the Saint Mary Cemetery.

This is the possible last resting place of the Lady in Grey, who has been
see walking on Noland Road in front of the property. The cemetery,
which is now owned by Independence Parks and Recreation, is the resting
place for some famous people including Confederate Brigadier General
John Taylor Hughes; James Andrew Liddil, member of the James Gang;
Samuel Locke Sayer, Judge of the 24th Judicial Circuit and member of
congress; Le Roy Smith, Jr, author of "The Fourth King"; and Samuel
Hughes Woodson, US Congressman.

Members of the QUEST team have visited this cemetery on several
occasions and have been greeted by ghosts almost every time. We have
seen several full and partial body apparitions, mysterious orbs darting
about, and mists in every part of the cemetery. One person who lives
nearby, says that they hear knocking on their back door, but when they go
to see who it is no one is there. His property borders the cemetery. Is a
ghost knocking on his door?

My daughter was searching one day for a particular marker at the site
belonging to her mother-in-law, when she sensed that someone was trying
to tell her to stop immediately and look down. She stopped and saw a
small marker for a baby girl who was born and died the same day. The

baby died on my daughter's exact same birth date, November 27, 1976. What made her look down and why there was a connection to my daughter is anyone's guess, but it seems to be an odd coincidence.

Pitcher Cemetery

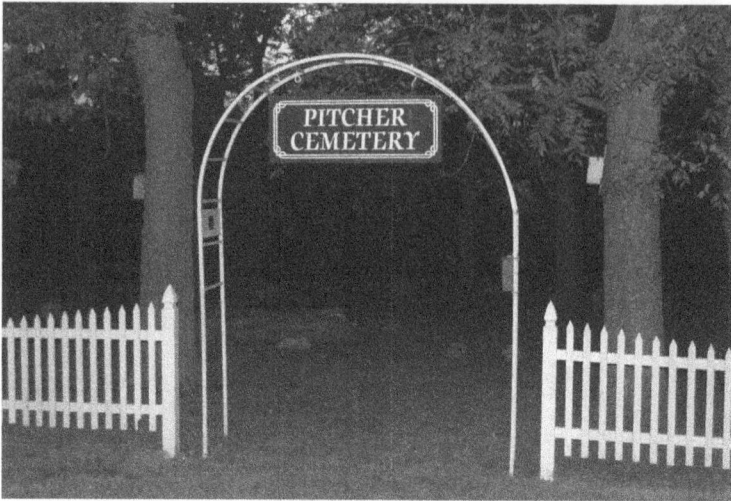

Located at Blue Ridge Blvd., North of Pitcher Road,
just north of 40 Highway.

This family burial plot for the Thomas Pitcher family established in 1830 is also the final resting place for men who fought in the Revolutionary and Civil Wars. Many Civil War soldiers who perished in battle nearby in 1864 are buried in a mass grave. Another mass grave is in the cemetery, which contains pioneers who perished from the Cholera epidemic of 1849—1851. Most of the grave markers are simple standing rocks without engraving. It is estimated that over 200 people are buried here without markers.

Visitors have reported hearing noises in the trees, and seeing glowing balls of light and floating apparitions.

I visited the site, which is right next to a small park with tables and a walking path, one evening at dusk and again a few days later with my daughter. As I approached the entrance, I felt the heavy presence of many souls who have not moved on. I went into a light trance and walked through the graveyard, camera in hand, taking photos while communicating with the dead. There are many spirits here who do not know they have died, or who are confused about where they are.

Unmarked gravestone at Pitcher – note the orb in upper right of photo

The 40 Highway Ghost

Photo: © iiievgeniy-Fotolia.com

In late December, 2002, Jamyi McLaughlin took some presents to a friend. She was driving in her car alone. Jamyi drove east on 40 Highway from I-70 and when she arrived at 40 Hwy and Manchester; she saw a two-car wreck with police and ambulance at the location. The traffic was slow and down to one lane. As Jamyi passed the accident, she noticed that a covered body was being taken out on a stretcher to the ambulance.

She glanced in the rear-view mirror and saw the image of a woman's ghost sitting in the back seat. The apparition had dark brown hair and glazed-over brown eyes. Jamyi was frightened at first, and knew immediately that it was the spirit of the woman from the accident. Jamyi is one of our Quest team members, so she is used to seeing and hearing very strange things, otherwise, I think she would have been even more unnerved.

Jamyi decided to ask the woman if she could help her. The woman nodded "yes" stiffly, and when she did so Jamyi noticed that the woman's neck was severely injured.

The woman then spoke out loud and said that she needed to get to her husband to tell him what happened and that "No one knows what happened." She said "Just drive straight ahead." *Note: It is very unusual for a ghost to interact with someone in this manner. I find it very interesting that she could be heard clearly while speaking.*

Jamyi, although still nervous, continued driving straight East on 40. When she got to a trailer park past Sutherland Lumber she saw the woman just step out of the car while the vehicle was moving, apparently headed for the trailer park.

A few days later, Jamyi and a friend looked in the obituaries in the newspaper and saw a photo of the woman who died on the date of the accident, and it was the same woman who appeared in Jamyi's car. During subsequent visits to the area she has not re-appeared, but if you drive on 40-Highway don't be surprised if a ghostly passenger suddenly appears in your back seat.

Old Independence Library Ghosts

228 & 230 W Maple Avenue, on the Square

This building is extremely haunted. It is located on the Independence Square is where two of the stores that I owned were located at the left and center of the photo. We rented the spaces. We started renovations in September of 2005 and worked on the stores until January and February until they were opened. The location also housed our Quest office, and our paranormal investigative team met here often to discuss cases. Everyone who spends any amount of time here quickly finds out that the place is very haunted.

This structure was built possibly around 1880, but no one has been able to get the full history of the building. It was once the Independence Library where Bess Truman worked and where Harry picked her up every day for lunch before they were married. The building has also housed a number of businesses in the past including Montgomery Wards. It has been divided in to several shops, but the exterior remains the same.

Needless to say, there has been a lot of activity at this location in the past.

During the period that we worked on the stores to get them ready we did some extensive remodeling. On a number of occasions my husband had electric power tools stop working even though there was no breaker thrown and the plugs were intact. The tools would start up again without anyone touching them. Sometimes when tools were left overnight they would be found in different locations the next day. We wondered if a spirit was playing tricks on us. Everyone working on the project heard disembodied voices on several occasions, but when I looked for a source I could not find anyone else in the building.

The voices continued throughout the entire three years we rented this retain and office space. The voices often fooled my assistant, our bookkeeper, the store manager, or one of our clerks into thinking that one of them said something. The voices were usually unintelligible, but on some instances could be clearly heard. Often a disembodied male voice would answer a question with a very clear "yes" or "no," or even with several words such as "I did it," or "It's over there."

The Lady in Red:

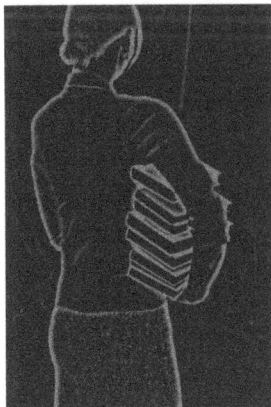

We had a camera set up with a monitor in the back office so we could see when someone entered the store. On several occasions I saw a woman dressed in a red skirt, white blouse with red trim on the collar, and red hat in what appeared to be 1930's clothing appear on the monitor. She has short, dark bobbed or pulled up hair and carries books in her left arm.

The lady walked determinedly through the middle of the stores, then through the center of the fireplace store through heavy cast iron stoves we have on display, then through the wall on the East side through a mantel. I believe that wall was an addition to the original building, so it may be that it was not there before she died. It seemed at times as if she is floating rather than walking, and at other times to walk normally.

Spirits have opened and closed the front doors on each of the stores. The bell sometimes goes off and the doors open, but there is no one there. These doors are very heavy solid wood and could not possibly open due to wind. We have seen some apparitions that look like solid people appear then disappear. This phenomenon occurs almost on a daily basis.

A child's voice was heard occasionally in the back room, or behind a closed locked door that leads to the storage room of the store on the other side. The child has been heard very late at night when all the stores are closed. We asked the other tenants if they have had children in their stores late at night, and they all said that they don't.

There have been several days when we saw someone walk in the store, but when we go to assist them there isn't anyone there. I believe this phenomenon is a *Time Imprint*, rather than a haunting. It can occur in places where lots of activity took place over a long period of time, which would certainly be the case here.

The spirits are especially active during October and November when the veil between the two worlds is thinnest. Sometimes it can get very annoying when you are trying to get your work done and get interrupted by the door bell, go out to greet a customer, and there is no one there – or you hear a person moving around and no one is around the area.

I taught a class on ghost hunting at this location in late October of 2006, and during the class a couple of the attendees noticed someone walking behind the curtain that leads to my office, and peeking out from behind it. No one was in there and there is no other way out. The attendees were quite excited to have seen a ghost appear during the class.

After the class, I spoke with Chris Brethwaite, a friend of mine who is a paranormal researcher. Chris was drawn to an area in front of a mantel display where he saw a woman appear. I did not see her, but felt a presence and saw a white mist. I walked to the area and we both noticed a very cold spot at that location. I took an EMF reading, and the meter spiked in that area. The spot warmed up after about 2 minutes.

A heavy Sony camera and case flew off of one of our bookshelves in December of 2006 while we had a lot of customers in both stores. I heard rustling of papers before it happened but was too busy to investigate. I got the feeling that the spirit was upset that we weren't paying attention to it. I asked it to quit the pranks and they stopped for a few days.

My assistant and I were talking over lunch one day and I told her about how my deceased grandfather, Boyd Kithcart, recently contacted me while I was trying to communicate with my other grandfather, Frank Lombardo. Right as I explained how he contacted me we both heard a violent shaking of metal objects next to the window in the office. I looked around the area and found no metal except for a heavy file cabinet. I don't know how that sound was made, but we both figured that Boyd was just letting us know that he is alive (in spirit) and well.

In July, 2007 we had another alarm go off at the store, and this was the fourth time it has been set off for no apparent reason. After three false alarms, the city charges a fee which of course, I didn't want to have to pay. The alarm company said that according to the sensors they have set up, that movement inside was setting it off, but we could find nothing in the stores that could move- except for ghosts. Two police officers met me at the store to examine the area. I finally told the officers after talking with them for 15 minutes that the only thing I could figure was ghostly activity, and to my surprise, they were not surprised at all. The officers told me that they get frequent paranormal calls in Independence. Now, I know there is a lot of haunting going on in this town, but for it to be so widespread that the police think nothing of it is amazing.

In October of 2007 I taught the "Haunted Independence" class again and we had three incidents. Several people told me they saw the heavy curtains I have in a doorway move on their own. These are too heavy to move due to air currents from the air conditioning. It is the same location where someone saw an apparition during class the previous year.

I saw the floating head of a man appear in front of one wall, and another woman saw it, too and screamed. She described it exactly as I saw it—with dark circles around his eyes and very creepy looking.

Next, a woman in the class said she saw a ghostly lady with blond hair wearing a white lace dress that was calf-length, and black patent leather shoes. The witness heard the ghost saying "look at me—I'm very fashionable." This is the exact same description that Chris Brethwaite gave me the previous year of a ghost that appeared to him in the same location. We felt her presence at the front of the store.

This is an example of how spirits may appear when you talk about them, as during the class on haunted sites in Independence. I guess it would be curiosity on their part—or maybe they are just letting us know that "yes, you are right, we are here." I noticed that there was more activity around when we started carrying Halloween costumes in 2007, which may have had something to do with the floating creepy head appearing at the same time. Prior to that there were no sightings of this particular ghost.

One pesky spirit seemed to hang around the back office where my bookkeeper, secretary, and assistant had desks. This male spirit would close and lock file cabinet drawers, talk out loud in answer to questions, and actually throw things across the room.

In the spring a tornado came through town and blew out one of the front display windows while I was closing up shop one day by myself. The tornado did not do any damage to any other building nearby. I ducked down behind the counter to get away from the flying glass and shelves, and after I thought it was safe I got up again to survey the damage. As I was standing there, a porcelain teapot came flying across the room in my direction, even though at that point the tornado and wind were long gone.

We decided that we'd had enough and moved out after our lease was up. There is only so much a person can take!

Since there are usually stores open at this location I imagine that you could visit these stores during the day while shopping.

Fort Osage

A few miles east of Independence
107 Osage Street, Sibley, MO 64088
816-650-5737
www.fortosagehs.com
Open March 1 through November 15 and Nov 16 – February 28

The second U.S. outpost built following the Louisiana Purchase, Fort Osage was a site chosen by William Clark in 1808. The Fort housed soldiers guarding the new territory, and acted as a stop for settlers heading west, but was abandoned by 1827. The current fort, which was reconstructed in the 1950's and 1960's, is an exact duplicate of the original.

Staff and visitors report sightings and interaction with ghostly soldiers in uniform, and Native Osage Indians in their natural dress. They also sense a presence of unseen people watching them and standing by, but when they turn around there is no one there. A man in military uniform, who speaks to people and poses for photos with visitors, does not in the photo when the picture later developed!

There are also reports of spoons falling to the floor, footsteps, spoons flying out of cooking pots, a hearth radiating a rainbow of colors, and flatware is often found turned upside-down in the dining hall. The staff does not recommend whistling, which may call in the "Woluska," who, according to Osage Indian legend, are tiny people who will cause you to have severe headaches. One staff member accidentally whistled one day, only to feel the immediate presence of another-worldly spirit.

Very near the fort is a cemetery that was established in 1810, where a group burial plot holds the remains of men who served at the fort and died of scurvy, and many other graves. Perhaps some of the haunting activity at the fort is due to spirits who linger at their graves nearby.

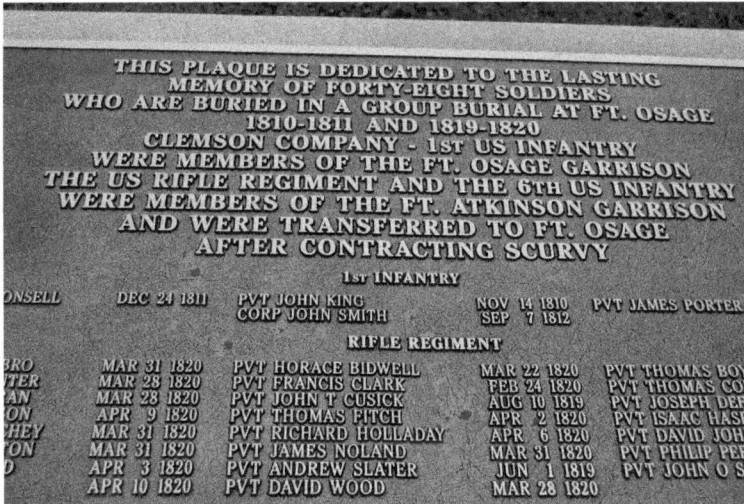

THIS PLAQUE IS DEDICATED TO THE LASTING
MEMORY OF FORTY-EIGHT SOLDIERS
WHO ARE BURIED IN A GROUP BURIAL AT FT. OSAGE
1810-1811 AND 1819-1820
CLEMSON COMPANY - 1st US INFANTRY
WERE MEMBERS OF THE FT. OSAGE GARRISON
THE US RIFLE REGIMENT AND THE 6TH US INFANTRY
WERE MEMBERS OF THE FT. ATKINSON GARRISON
AND WERE TRANSFERRED TO FT. OSAGE
AFTER CONTRACTING SCURVY

1st INFANTRY

| ONSELL | DEC 24 1811 | PVT JOHN KING | NOV 14 1810 | PVT JAMES PORTER |
| | | CORP JOHN SMITH | SEP 7 1812 | |

RIFLE REGIMENT

BRO	MAR 31 1820	PVT HORACE BIDWELL	MAR 22 1820	PVT THOMAS BOY
TER	MAR 28 1820	PVT FRANCIS CLARK	FEB 24 1820	PVT THOMAS CO
AN	MAR 28 1820	PVT JOHN T CUSICK	AUG 10 1819	PVT JOSEPH DEF
ON	APR 9 1820	PVT THOMAS FITCH	APR 2 1820	PVT ISAAC HASE
HEY	MAR 31 1820	PVT RICHARD HOLLADAY	APR 6 1820	PVT DAVID JOH
ON	MAR 31 1820	PVT JAMES NOLAND	MAR 31 1820	PVT PHILIP PER
D	APR 3 1820	PVT ANDREW SLATER	JUN 1 1819	PVT JOHN O S
	APR 10 1820	PVT DAVID WOOD	MAR 28 1820	

Photos of the cemetery at Fort Osage

Missouri Town 1855

8010 E Park Road
Lee's Summit, Missouri in Fleming Park
Open March 1—November 15 daily.
November 16—February 29 Weekends only
816-503-4860 www.jacksongov.org

Each of the twenty-five buildings here dating from 1820-1860 have been moved from other sites, and carefully reassembled at this living history museum. Craftsmen, business owners, farmers, and homesteaders work daily in period attire.

Legend has it that in one of the cabins a little girl was bitten by a snake and died. In this cabin strange smells and noises have been noted by staff and visitors.

Some visitors have seen a candle burning in the window of the upper floor, which goes out when a person arrives in the room, leaving the smoking candle. On one occasion a volunteer blew out one candle on a table on the first floor, and all the rest of the candles in the room went out at the same time.

Some of the staff members report hearing loud knocking on the front door, but when they open the door there is no one there. A broom near the fireplace has been seen flying across the room by itself. This is no doubt, a very haunted house. The house gained the attention of local news and has been featured on TV for years.

The 1859 Jail

217 N Main Street, Independence, MO
www.jchs.org/jail/museum.html
Open year-round to the public

In January of 2006 my mother and father in-law visited my husband and I and we all toured the 1859 Jail, Marshal's home and Museum run by the Jackson County Historical Society. A couple of the more famous "guests" who stayed here were Frank James and William Clark Quantrill. Prior to the Civil War, a person could end up in jail for firing guns in town, operating a gaming house, disturbing the peace, disturbing a religious meeting, horse racing on public streets, or building a privy "not over a pit."

It was our first visit to this building. Immediately upon entering the inmate area I felt the presence of several spirits. This did not surprise me at all considering the age and history of the building, but I was taken somewhat aback by the heaviness of the air. I knew that all of the spirits were not benign. As we walked past the 12 cold limestone cells, I could feel the loneliness and despair of the inmates from long ago. I could have looked at the past to see what happened then, but tried to shake off that feeling and see what I could tell about any lingering spirits instead.

William Quantrill
Public Domain photo

I could see several people inside one tiny cell because there were not enough rooms for all of the inmates in the 12 cells. Some of the people had done nothing wrong and felt greatly wronged by their accusers. I heard crying that sounded like children or young women, and saw a pregnant woman. Later I found out that women and children who were "on the wrong side" were kept in the jail just before and during the Civil War.

Upon entering Frank James' old cell I immediately felt his presence and personality. Knowing nothing of Frank at the time, I was surprised to sense that he was intelligent, could read and write quite well, and was obviously well–educated and well-mannered. It seemed that he could come and go as he pleased and his cell door was open. Surprisingly, I did not sense that he was really a bad guy, or that he thought he was bad at all, which was not exactly what I was expecting from an outlaw. When we toured the museum later I found that my assessment of Frank was exactly right. Frank was friends with the Marshal and usually had dinner in their house, located right above the jail in the same building during his "incarceration."

While still in Frank's cell I sensed a different and more sinister spirit.

This is a lingering ghost who stays behind and lives in that cell or near it. The spirit is a tall heavy-set Caucasian man with dark long hair that he pulls back. He also wears a long beard. He feels that he is trapped in this location forever and he is somewhat insane. I believe that he takes energy from people who visit and that keep his spirit going. I would not attempt to confront him as he is extremely negative.

I then saw a tall African-American man incarcerated here who has a strong personality, was a murderer, and was afraid of no one in his lifetime. I also saw a fight between two prisoners who were in cells across from each other.

Interestingly, a lady and her daughter came into my shop one day and related a story about how during their visit to the jail someone grabbed the daughter's arm, but when she looked around there was no one there. Then she saw two men in Civil war uniforms (one Union, one Confederate) fighting and shaking their fists at each other through the cell bars. One man was released from his cell and the guard handed him his guns. The man then turned and shot and killed the man he was arguing with right in front of the guard. The woman was so frightened by this sight that she will not return for another visit.

We toured the Marshal's home and when we got to the large bedroom area upstairs I could feel the presence of a strong woman. She does not want to leave and is very attached to the home. She feels that it is her home and all the visitors that go through there now are intruders. The woman does walk around to other parts of the building, including the gift shop area, and lets people know she is there. I hear the name of Cassie. It is speculated that this spirit is the wife of Marshall Jim Knowles.

Some people feel that the adjoining Marshal's house is haunted by Jim Knowles, who was killed trying to settle fight between two prisoners; one a Northern sympathizer, and one a Southern Sympathizer. Others say that a deputy marshal who was killed in June of 1866 during a jailbreak haunts the jail.

The jail has seen a lot of inmates come and go, including women and children, during the pre-civil war hostilities as well as during the war and later. Each 6' x 9' cell was designed to hold three prisoners, but often as many as 20 prisoners were placed in the cells. Due to overcrowding, several other buildings were used to house prisoners. One of these buildings collapsed, killing several young girls. After this, Tom Ewing issued Order No 11 in 1863, which forced anyone who sympathized or dealt with the Confederacy to leave Missouri. Some who didn't obey ended up in this jail.

The jail staff and other guests have related many unusual experiences in the jail, and have seen apparitions of men, women, and children, sounds of children playing and crying, whispering, sounds of people moving around in the jail area when no one is there, moaning, and items being moved. Sometimes the radio even turns off and on by itself.

There is definitely an ominous feeling about this place, I would not recommend taking small children here, but it is a great place for a real scare if you're up for it!

The jail has been restored and is open to the public from March through October.

Note: every year during October, the Independence Square Association hosts ghost tours on Friday nights, and the jail is one of the stops. A paranormal group host tours at the jail. Visit www.theindependencesquare.com for more information.

Haunted Rotary Park

10615 E 24th St. S, Independence, MO
Rotary Park is open to the public year-round

Rotary Park is located one block south of 23rd Street, and a few blocks
west of Sterling. It is a popular place for children to play and for adults to
walk on the track around the park. There is a creek on the South and West
side of the park and a big hill on the West side beyond the creek. The park
is open to the public year-round. What most visitors don't realize is that
the living aren't the only ones there.

My eldest daughter and I used to take walks in the park almost nightly
starting in the spring of 2003. We did not notice anything unusual about it
until September, 2003 when the park cleared out earlier in the evenings and
was fairly quiet. Often, we were the only two people walking.

One evening while briskly walking along on the east side of the track, I felt a shaking in my entire body that started at my head and jaw, then traveled downward. The shaking stopped when I moved forward. I stepped backward a few steps and the shaking started again. I could feel a definite difference in the air in this one spot. My daughter stepped into the space and said she felt an "electrical" type feeling.

The next evening, we returned to the park again. The spot was still there and felt the same to me. I was intrigued about it and decided to go into trance to see what it may be. We sat on a park bench, and suddenly I saw spirits all around the park. Some were Imprints, which are what I call *Time Imprints,* where a psychic can see events that happened in the past as if watching a movie. I saw small log cabins, several tent camps, lots of camp fires and many people milling about. And a few of the apparitions were ghosts.

The first ghost is a man who hangs out around the south edge of the park, and is there every time I visit. He looks like an old prospector and has a gold pan in his hand. He wears a dirty white shirt, suspenders, pants, boots and an old raggedy hat. I see a tent set up with a fire and pot in front of the tent facing the stream. The man means no one any harm, but he does not like it when people get too close to him. He seems not to realize that he is dead. There is an odd round impression in the grass in this area where the grass blades move in the opposite direction of the wind at times. My daughter and I have noted this on several occasions. I believe that is where the man's tent was. The name he gave me is "Old George."

Another spirit that haunts this park is that of a young woman in her late 20's or early 30's. She wears a long dress and stands next to a large tree on the west center side of the park near the walking track. She told me her name was *Annie.* She feels much attached to the location. When I did a reading of the area, I also got the name of Frank James but did not know why there was a relationship to this woman at the time of my first encounter with her.

I asked local history buff Joe Rudzik about this location and he told me that where there is currently a brick schoolhouse at the top of the hill is the exact spot where there was an older school. Frank James' girlfriend was the schoolteacher. In those days it was customary for a teacher to be unmarried. When the community refused to let Ann marry Frank James and still retain her position as teacher, the school mysteriously burned down. The entire town thought it was Frank who was responsible, but couldn't prove it.

Frank then offered to rebuild the school, but only if his new wife could continue to teach there. The town council agreed, and the new school was built. I think this woman is the ghost of Frank James' wife. I did research on the internet and found that her name was Ann "Annie" Ralston (1853-1944). Frank James (1843-1915) is buried in Hill Park, directly across 23rd Street, and I Ann is buried with him. I believe her ghost haunts the park because she has a strong attachment to the school.

This is a good place to visit if you want to see a ghost. If you place your hand on the east side of the tree 12" away from the tree approximately 5' up from the ground it will feel noticeably warmer than the air around it. *Note: most ghosts make the area around them colder, not warmer, but in this case, it is the opposite of what you would expect.*

We took a tour to this park on Halloween night in 2004, and I saw a woman materialize there and then turn to white mist. I immediately took the group to the tree, and all eleven people could feel the warm spot next to the tree. Three people could see a white mist in the same space.

Investigators touching the warm spot on the "Annie" tree. This measured 9 degrees warmer than the ambient air and rest of the tree.

In October of 2007, I took a class outing to this park on two different nights. On both occasions everyone felt a difference in temperature next to the "Anne" tree and I was able to get several photos of orbs. One class attendee had is video camera battery suddenly drain power, from 1 hour of time left to 45 seconds left in an instant. He was standing near the tree when it happened. Also, at the same moment, I saw Anne come out of the tree. Perhaps she could not materialize until she had taken some energy from the battery.

Other spirits appear in the form of tiny bright orbs and flit about the tree tops. They occasionally wander down to the center of the trees. These feel like young children or possibly sprites (young, fairy-like creatures).

There are tree spirits in at least three locations in the park. Tree spirits are actual spirits of the trees and they have very individual personalities. They are visible to almost everyone who looks at them. The tree spirits have eyes, nose and mouth, and their face changes shape and may even speak to you if you look at them for a few minutes.

If you go to the park, look for knots in the trees and watch them for a few minutes. If you listen quietly, they may tell you something about things they have seen in the past. This is a good way to do an investigation of an older area and get help with finding out what may have transpired there.

On the west side in the trees on the other side of the stream there is a dark, very ominous heavy presence. On some nights it feels as if something is watching from that location. One evening in early 2005 a dark shape moved forward to the near side of the stream. We could not determine what it was - if it was a ghost or other entity. This location is not far from the school on the hill, which is also haunted and is mentioned later in this book. Later visits proved that the dark energy has not left the area. I don't recommend crossing the stream to the other side.

Rotary Park Orbs

The photo above was taken by one of my assistants as she followed me around the park. The brightness is increased in order to see the orbs better. At this particular spot I had just requested that any spirits that may be present come forward and communicate with us. Note multiple orbs in the picture.

The night was very still, so I don't believe that the orbs were caused by dust particles. Also, the camera lens was clean and we saw nothing in the photographs taken before or after this one.

Above: a light and an orb appear behind me at the park.

Bingham-Waggoner Estate

313 West Pacific
Independence, MO 64055
816-461-3491
www.bwestate.org
Open April through October to the public, available for special events.

This estate was built in 1855 and was owned by the famous artist George Caleb Bingham, and later by the Waggoner family, founders of Waggoner Gates Mill. The site is now used for special events and weddings. It is open to the public for tours and events.

I visited the site for a Chamber of Commerce function one evening and as I walked by the landing below the staircase, I was drawn to stop and look at it. I was very surprised to see a transparent woman in a bridal gown lying on the landing. She wasn't moving. I asked one of the employees if they had had any ghostly activity at the site and she told me that the ghost of a bride sometimes appears near the staircase, where she tripped on her wedding gown and fell to her death sometime in the late 1970's.

On a later date, I was walking around the grounds of the estate when I got the eerie feeling of being watched. I was with one of our team members and mentioned this to her. She said that she felt some type of energy coming from the barn area, and we walked in that direction, expecting to see a person working there. There was no one in the barn and it was dark. The feeling grew more intense, as if something was pushing us away. A dark energy approached us – this was not only a feeling but also something we saw as well. We both backed up and the sensation stopped, however, the feeling of being watched never left us. I have not returned to the site since.

Photos of orbs have been taken in the bedrooms and near the stairs by visitors to the estate.

The barn at the Bingham-Waggoner Estate

Photo:
Melissa M Kothe

Wikimedia Commons

The Ginger House Museum

Birthplace of a famous actress

100 W Moore Street
Independence, Missouri
Open April 1 – Sept 30 1- 6 pm Wed – Sat
Open most weekends in the Fall And for private group tours and paranormal groups year-round

This 1906 Craftsman Bungalow is the birthplace of a famed actress/dancer/singer. We can't mention her name here due to an agreement with her estate. The actress started performing at a very young age. She first posed for a newspaper ad at age 18 months, and performed at the Kansas City Club at age four. She and her mother lived in this house for four or five years before she went to live with her grandparents in Kansas City until age 9, when she moved to Texas with her mother and new stepfather, John Rogers.

The actress won the Texas State Charleston Contest, which rocketed her into Vaudeville, then Broadway, where at age 19 she became an instant star. She made 73 films, starred on Broadway and the London stage, and in her own productions. She wrote an autobiography and tells the story of her birth and how her mother, at age 19 and abandoned by her husband, rented this house and worked nearby.

She and her mother visited the house on several occasions. She visited the family that lived in the house and did a photo shoot for LIFE Magazine in 1942. She again visited in 1964 for "a special day designated by president Harry Truman. who presided over the festivities, and again for a special day hosted by the City in 1994. At that time, the property was declared an Historic Landmark Property by the City of Independence.

My husband and I own a real estate investment company and purchased this house in February of 2016 with the intention of restoring it and turning it into a museum. I have long been a history buff and fan of the actress, as was my mother. We knew that the house existed, but had never visited the site. I thought that this landmark should be preserved for future generations to enjoy. The furthest thing from my mind was that the house was haunted. But haunted it is.

One day, not long after purchasing the house, my husband told me that he heard unexplained noises and that lights were going off and on. He said that something must be wrong with the wiring. We soon had a licensed electrician out to remove all the old wiring and install new wiring and a service entry panel, so we thought that would be end of the problem. It wasn't. The lights continue to flicker on and off, and we hear unexplained noises, especially if there is anyone in the house investigating paranormal activity or if we talk about the actress.

One evening, I drove by the house and saw a figure standing next to the post to the left of the door. The figure was misty and transparent, but I could clearly see that it was the actress wearing a skirt. She looked to be about 30 years old. I was very shocked to see this – after all, she only lived in the house for a few years. But she did visit the site several times, so perhaps she does have an attachment to the property. Could she have been the one who caused the lights to flicker and did she make the strange noises? I dismissed this and didn't even mention it to my husband, but several days later I would get a strange phone call that would change everything I thought I knew about the property.

Being a paranormal investigator myself, I should have done a thorough investigation of the property before getting into the renovation work, but my mind was not on ghosts, I was focused on the historic value of the home and its importance as the birthplace of the actress.

My business phone rang and my secretary said the call was for me. It was Stephanie Turbeville with KC Paranormal, and she had a story to tell me. She was driving home after an investigation one night when she felt compelled to drive by the Ginger House. She got out of her car and stood on the sidewalk, when to her amazement, she saw the actress appear on the porch standing next to the pillar at the left of the stairs. The apparition smiled at Stephanie. I was shocked to hear that someone else saw the spirit standing in the exact same place where I saw her – and without knowing anything about my experience. At that point, I decided that the best thing to do would be to have a different paranormal team investigate the property in order to get an unbiased opinion about the activity going on there.

While waiting for Stephanie and her partner, Jan, to come out I stood in the house one evening in the room where the actress was born. I was trying to decide how to finish the look of the room, when suddenly a foggy shape began to appear in front of the wall. I just stood there, waiting to see what would happen, when the shape formed into a woman wearing a long ballgown. It was the actress, and she looked like she was in her 30's. The mist dissipated quickly. Now I couldn't wait to do a full investigation and try to get evidence on film or audio!

Stephanie and Jan arrived a few days later, and my brother went with me for the investigation. I brought my FLIR camera, cell phone, and a good still camera, and Stephanie brought several night vision video cameras, flashlights, audio recorders, an EMF meter, and a spirit box. A spirit box creates white noise, which sometimes allows spirits to communicate with us from the other side. None of the four of us were prepared for what would happen that night.

The house had no furniture in it, so I brought four folding chairs with me. Other than the chairs and the equipment, there was nothing in the house, and no one else around. I took a good look around the perimeter to make sure no one was nearby. Certainly, no one was in the basement.

We set up in the living room. It was approximately 9:00 pm and dark outside. A few minutes into the investigation, all of us began to see white and blue balls of light, or orbs, moving about the house in all of the rooms. These objects seemed to be moving intelligently, first heading one direction, then another, as if they were scanning the house. Some went through walls or windows. Many were visible with the naked eye, but when looking through a night vision camera, we were able to see many more. Next, we began to hear noises that seemed to be coming from the basement. The door was locked, so no one could have been in the basement. The noises then moved to the kitchen. It sounded like banging and like someone with heavy boots was walking on the stairs and floor. None of us could explain it and we wondered if we were perhaps dealing with a spirit other than the one we thought.

A purple ball of light the size of a baseball came through the front door and headed directly for my brother, who was in the corner opposite from the door. My brother put his finger up and the orb attached itself to the end of his finger, approximately 1/2" away. As he moved his hand back and forth, the orb followed and "stuck" to him. I then put my finger next to it, and it attached itself to my finger and followed as I moved around. It then detached, and flew straight out the front door again! All four of us witnessed this- and all of us were suitable surprised and dumbfounded. Obviously, the orb had intelligence, but what it was is anyone's guess.

Stephanie wanted to try the spirit box and turned it on. She asked questions such as "Is anyone here with us?" and "Why are you here?" We all asked questions at different times, and someone answered us through the spirit box. We heard three male voices and two female voices. One of the female voices we all recognized – it was the actress! The other sounded older, and was likely her mother. They answered our questions, sometimes one word was spoken, sometimes a sentence could be heard. Their voices sounded like they were coming through a tube, but were clear most of the time. We all came away with the impression that the actress and her mother were happy with what we were doing with the house and that they were going to help us with the project. One of my big questions was "Will the City approve our plans for a museum?" to which a voice answered "Yes they will." And indeed a few months later they did. I am certain that the ghost either knew about the approval in advance or had something to do with getting the approval.

The male voices, however, were more ominous. We heard the words "Murder," and "Basement," among others. That could have been the reason for the noises coming from the basement. The basement floor is made of dirt. Stephanie looked at me as if to say we should dig it up, and I said out loud, "No, I'm not going to let anyone dig it up to see if there is a body down there!

After a while things started to slow down. The two women indicated that they were being blocked from speaking to us by others. All became quiet and no more answers came through the spirit box so Jan turned it off. Next, he placed a flashlight on the floor as a means of communication. He asked "Is _____ here with us?" and the flashlight came on. He asked her to turn it off, and it went off! This happened several times as each of us asked a question, and it was clear that the actress was able to communicate via the flashlight.

Now here is where things got really, really weird. I looked into the bedroom off of the dining room and saw an irregular blue/black glowing web-like structure in the doorway. In the web were faces – but not humans – they looked like extraterrestrials! I said nothing as I wanted to see if anyone else noticed them, and soon my brother, who does not often see anything unusual, looked in that direction and said "I think it is time to leave." We asked why and he said "There are Aliens in that doorway." Stephanie and Jan moved to where they could see the doorway and they saw the same thing. There were several odd-looking faces and partial bodies of what looked like Alien Grays. This was certainly the very last thing we were expecting!

We all decided that we'd had enough and packed up to go home. Stephanie and I both agreed that we got more evidence of paranormal activity in this house in just 2 ½ hours than any other investigation either of us had done before. As of the writing of this book, workers at the museum have reported hearing strange sounds in the house during the day when things a quiet. When they investigate, they can find no explanation. We also have experienced issues with the alarm system and cameras that can't be explained. I believe strongly that the renovation process brought forth some ghosts from the past.

In the fall of 2018 I purchased a number of antique items for the house from a local woman who is selling part of her parent's estate. Her father was a furniture maker and had a child's rocker in his inventory which he had repaired for a client many years ago, but they never picked it up. I thought the rocker, along with some antique toys, would be a good addition to the museum. But I wasn't prepared to bring a ghost along with the rocker.

During an investigation in October of 2018 with six other persons, we found that as each person walked by the small rocker, they made a comment about it. Some people asked where it came from, others said they got a strange feeling about it. Since the rocker had only been in the museum for a couple of days and I hadn't noticed anything odd about it, this struck me as strange. The people walking through the house focused on that chair. I asked everyone to run their EMF meter by the chair as they walked through the rooms. Each one got high readings over the chair, but no where else. In fact, the house was eerily quiet the entire evening – as it had never been before.

While watching the chair from another room the apparition of a little girl around age four appeared sitting in the chair, then moved across the room and dissipated. At that point, I knew something was going on, but don't know if the chair brought a spirit with it or if the spirit of a child that had lived in the house was sitting in the chair.

Finally, after a couple of hours I decided to move the Rem Pod to the chair. A Rem Pod is a very sensitive instrument that detects changes in electro-magnetic field and senses static electricity. Almost immediately it started to beep and the lights flashed. We watched in amazement, as the beeps got more frequent, until we all looked at each other and said "Morse Code" at the same time. The beeps were a series of long and short sounds which reminded us of Morse Code. This kept up for a full 30 minutes, until I finally turned off the equipment.

I've invited a couple of other paranormal investigators to come to the museum to get their opinions about the situation. I will keep posting about this site on my blog *Haunted Missouri Sites* at http://www.hauntedmissourisites.blogspot.com.

Longview Mansion and Farm

Photo: Sharonclay022 – Wikimedia Commons
3361 SW Longview Road
Lee's Summit, MO 64081
www.longviewmansion.com   816-761-6669
Weekday tours and luncheons for 20 people or more. Available for
weddings and special events. Not available for individual tours.

Longview Farm was built in 1914 by millionaire lumberman Robert A.
Long (1850-1934), who was married and had two girls. The farm had 51
buildings, a horse arena, racetrack, greenhouses, a school and church. Up
to 200 people lived and worked on the farm at one time, but during
construction approximately 2,000 people worked there.

Loula Long-Combs (1881-1971), Robert Long's daughter was an
internationally renowned horsewoman. She was the first woman to drive in
competition at Madison Square Garden, where she was inducted into the
Madison Square Garden Hall of Fame. She was known as the "Queen of

the American Royal," because of her many prizes won at that event.

All of Loula's prize winning hours, including her favorite "Revelation," are buried in front of the old show horse arena, which is now Longview Farm Elementary School. Both Robert and his daughter Loula, who operated the farm after her father's death until 1971, died in the mansion.

The mansion has 48 rooms, six fireplaces, 14 bedrooms and 10 baths. It also has the first central vacuum system west of the Mississippi.

The ghost tales include multiple sightings of Loula on her horse in the indoor ring, sounds of hoof beats when no horse is near, unexplained fog indoors and out, and apparitions of a woman riding a horse around the property. Staff reports that Loula's bed often appears to be slept in, even though no one sleeps in this bed. Apparently, Loula loved this place and her horse so much that she is still here long after her death.

During one of my visits to this location during the Holidays I saw a man dressed in 1930's attire standing in the back corner of what used to be one of the main barns, watching the activities. At first I thought he was part of the festivities since there were several people in costume, but when I glanced back he was gone. Perhaps it was just Robert Long checking on his property.

George Brett Bridge

George Brett Bridge is located at I-70 highway and Blue Ridge by the Royals Stadium. The bridge was rebuilt in 2011 after it had fallen into serious disrepair. The area is extremely busy with lots of traffic and it is dangerous to walk across the bridge.

George Brett is a retired major league baseball player, famous for playing third base and being a heavy hitter. He is beloved by Kansas City Royals fans.

Multiple people have seen the ghost of a man who was killed by a drunk driver, supposedly on the night before or the night the Royal's won the World Series against the Cardinals in 1985. The man has been seen standing and walking on and near the bridge after Royals games. He wears his Royals baseball cap.

Part II
Haunted Places
In Independence
-not open to the public

The Haunted Dairy Farm

On Crenshaw Road near the Little Blue River
Private property

One of the most famous legends in the area, this site is purported to be very haunted. Witnesses have reported seeing the full-body apparition of an elderly man, dark shadows, fast moving objects, and many have seen a man covered in a gray sheet, yet he disappears in thin air. One entity seems very unfriendly and aggressive. Passersby have seen ghosts from their vehicles.

We can't give an address because the owners don't want visitors.

Nova Center

Private school for the handicapped
Not open to the public

The site of this school is the original location of an early Independence public school. This site is where Ann Ralston, wife of Frank James, taught class. The city did not want her to continue teaching after her marriage to Frank, as was the custom at the time. However, the school mysteriously burned down and Frank James offered to rebuild it as long as the rules were changed so that Ann could keep teaching after they were married. The city agreed, and Frank built a new school for the city.

    A dark energy is seen and felt on the hill behind this property, which is adjacent to Rotary Park (see the chapter on Rotary Park). Lights have been seen turning off and on inside the building late at night when no one should be there. The photo above was taken at dusk and there are several tiny pinpoints of white light on the far right side of the photo, which I can find no explanation.

Photo taken on site, strange fog appeared in front of the camera

In 2006 I visited the site with my daughter, who is another investigator on our team. The place was completely dark and looked abandoned. As we stood in front of the building, a light came on in the circular window above the main entrance and a shadow walked by the light. We felt a dark, ominous presence, similar to the one on the hill behind the school, and decided to leave the area very quickly. The presence is still there.

The photo above was taken outside my vehicle, as were the other photos shown here. This photo should have been as clear as the one on the next page below, but it is not. It looks somewhat distorted, and there are foggy objects and dark objects that I can't explain or find a source for. If this had been taken through a car window, reflections might have been the explanations for some of the objects, which were not visible at the time the picture was taken.

A picture of the front of Nova Center taken at dusk. A couple of "orbs" are in the photo (see arrows).

My Haunted House

She speaks in muted tones
appears as a wisp of smoke
about the house she roams
among the living folk
By Margie Kay

Naturally I can't print the address here or we'd have tourists stopping by every day— but my home is located in an older section of Independence. The house is 2,500 square feet and consists of an older portion that was built in 1929 and a newer, larger addition that was built in 1986. We did extensive remodeling of the house from 1987 to 1995 which included removal of walls and relocation of the lower level bathroom, the addition of a dormer, and expansion of the kitchen. The original floor plan is pretty much the same, not countin the addition. However, the renovations seemed to stir the ghostly occupants of the house, which is not an uncommon occurrence. Ghosts don't seem to like changes.

I should mention that every house I have lived in has been haunted. When this happens, many people say the *person* is haunted, rather than the houses but I believe that it just means I'm sensitive to ghosts that are always there. I am certain that most older houses are haunted, and that because I am psychic I see the spirits, whereas someone who is not as in tune with the higher dimensions may never see a ghost, even though spirits are around all the time. I'll talk about this more in the last chapter of the book.

From the time we moved in until the year 2000, my husband and I operated a service business out of our home and had a secretary come in every weekday. The older front living room was converted to an office. My two daughters, ages ten and eight, first brought ghostly activity to my attention when they reported hearing footsteps on the staircase to their bedroom. Sure enough, we heard a man's heavy stomping and a woman's lighter steps on numerous occasions. When people visiting heard the steps and asked about it we said "Oh, that's just one of the ghosts." I decided to do a "scan" on the house to see if I could pick up anything else, and I sensed the presence of several adults and a little girl. None of them felt threatening, so we decided to leave the ghosts alone.

In September of 1988 the paranormal activity picked up. During a Girl Scout sleepover at our house the ghosts apparently decided to have some fun scaring the kids. They made noises on the stairs, in the walls and in the basement under the front office where most of the girls were sleeping. None of the girls slept very well and they couldn't wait to leave in the morning.

The next morning five girls were still in the living room/office when I asked my oldest daughter where my keys were after I had looked for them for several minutes unsuccessfully. She and I were standing about three feet apart. The keys lifted up off of a file cabinet, floated in the air a distance of five feet to a spot right between us at eye level, then dropped to the floor. All of the girl scouts saw this happen and ran - screaming—out of the house. They waited outside for their parents to pick them up. Of course, none of them would visit again. Now I knew I was dealing with at least one prankster spirit and did not appreciate it.

In January of 1989 I hired a new secretary (the first one quit due to paranormal activity while I was gone) and told her about the ghostly activity that had happened in the past but that it was pretty quiet now. She did not seem alarmed. I felt that I should tell her about it up front so she wouldn't be afraid if anything did happen and hopefully would not quit as my previous assistant had.

Not more than two weeks later I had to leave to do errands and came back to find my secretary wide-eyed and packing up her things. When I inquired as to why she was leaving, she explained that she heard a noise in the basement and called out to see if it was my husband, Gene. When no one answered, she thought that he might be hurt and went downstairs to investigate. No one replied, but she did see a little girl (about age 11) with long blond hair in a long white nightgown standing by the furnace. When she asked if the little girl was ok, the girl slowly faded away and disappeared. I could not convince my new secretary to stay after that incident and had to find yet another assistant.

I did a scan of the house and saw the little girl. She told me that she died in the house and it was because of the furnace. I did research but could find no reference to this incident. The only thing I could figure was that it may have been Carbon Monoxide poisoning due to a faulty heat exchanger or blocked flue.

I asked the neighbors in the area if they knew anything about the history of the house and found out that there were 11 people living in the house just prior to our moving in. Since the room addition was incomplete, there was no heat or air conditioning in that part of the house so they must have all lived in the older, smaller side. That would have been very crowded and stressful. The neighbors told us that they heard a lot of fighting and yelling on several occasions, and that one or two of the men used to stomp up and down the stairs quite often. The neighbors also told me that a young girl was found dead in the house in the 1960's or 70's but they didn't remember the cause of death.

In the summer of 1990 one of the ghosts became audible. I was cleaning up the kitchen while my daughter, Maria, and her friend Jennifer, who was spending the night, washed and dried the dishes. Both girls were 12 years old at the time. Our house had not yet been remodeled in the kitchen and the old original cabinets were still in place. The bottom drawer often stuck so I told my family to always leave it open slightly so we could get it open more easily. On this occasion, I went to the drawer and it was tightly closed, so I said, "Alright, who closed this drawer?" To which a man's deep voice replied, "I did." The girls and I looked at each other, dumbfounded. We questioned each other as to who made that remark, and we decided that I would ask the question again while we watched each other. I asked, and this time the ghost answered in a very annoyed manner. The reply was a louder male voice saying "I DID!" That was the last time that Jennifer visited us. I wish I had that on tape.

Not long after the kitchen episode, a ghost appeared in the window one evening. I drove up to the house after an outing with my daughters, and as we pulled up to the side of the house a man pulled back the drapes and looked out of the window at us. We thought it was a burglar and went next door to call the police, who found nothing when they arrived. The doors and windows were all locked. My daughters and I stared at each other, knowing immediately that it was a manifestation of one of our spirits.

In 1991 I finally did a clearing of the house because one of the male spirits seemed to be gaining in strength and negativity. He did things to scare my daughters such as morphing posters of rock stars on the walls in my oldest daughter's room into hideous creatures. At first, I was skeptical about it, but she convinced me to watch it for myself.

I sat on her bed and watched the poster one evening, and after only a few minutes could see the morphing of the poster. I removed the poster and replaced it with another, friendlier poster of another rock star, but the same thing happened a few days later. She also said she saw a dark figure floating around the room at night, which looked blacker than the darkened room. I could tell that she was afraid, which was odd because she normally is not spooked by anything.

I began to see evidence of something draining energy from both of my daughters, because they seemed tired all the time. I suspected that it was the male entity. My daughters had not yet learned how to protect themselves fully against psychic attack, and it was evident that something needed to be done.

I did a clearing of the entire house, starting in the girl's rooms. (See the chapter on how to get rid of spirits at the end of the book or in my book Gateway to the Dead" to see the techniques I used. Suffice it to say that the negative spirit stopped his activities and we have had no evidence of his return since.

The little girl has since appeared to us. She starts out as a white mist, and then materializes fully except for her feet. She always wears a white long nightgown and floats in the air. The spirit can answer us by closing the basement door or knocking on it or something else in the basement. We feel her presence often. I have asked her if she would like to go to the light and get on with the next world, but as of 2018 she is not ready to go yet. It seems that she has a strong attachment to the house. Since she is doing no one any harm I will leave her alone for now.

The Portal

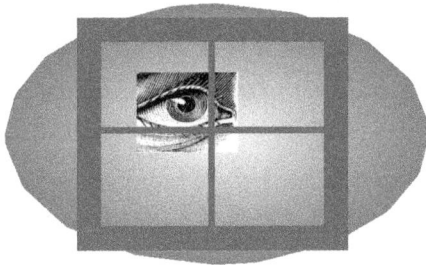

We have a portal at our house located inside and outside the house on one exterior wall off of the deck. This portal appears at times in the shape of a round swirling hole and at other times as a four-pane window that some psychics can see. Others don't see it but can feel it. We suspect that this is an inter-dimensional gateway that not only spirits travel through, but other unknown entities as well.

I was surprised to find that this type of four-pane portal appears to other people as well, and has been photographed. We have not been able to get any photos of our portal yet – every time we try nothing shows up on in the picture. It appears open more and more frequently as time goes by, and is usually open during the last three months of the year

During the time that this portal is open we see spirits appear to go in and out of it. Often, these entities don't seem to notice me or anyone else and just pass through the portal and move through our yard to go on about their business somewhere. Sometimes, whatever comes to our world does seem to be observing me and/or my family. Why the portal is in this particular location, when it first appeared, or why it is there are all questions that remain unanswered. Maybe it has always been there—even before the house was built. I suspect that the area, which is about a mile away from the RLDS Temple, may be part of a larger portal or vortex.

The Temple location is very near the spot where Joseph Smith wanted it to be built. He said that Independence was the new Zion. Did he maybe know something about vortexes? I have observed the portal on several occasions and it may be that this is how beings and spirits from other dimensions travel back and forth from their dimension to ours.

September 1987: My husband and I were gone for a day and my two teenage daughters stayed behind. They had two girlfriends stay the night. My oldest child, Mia, saw a large face appear on the wall next to the staircase in the new addition of the house. She pointed it out to the other girls and they all saw it as well. The face was three-dimensional, and measured about 3' in height. It seemed to be looking at them. Mia said it appeared more alien than human but was not sure why it was there or what it was. The face has not reappeared since, but a silvery mist has appeared in the same location several times. I suspect that whatever it was it came through the portal, which is approximately 15' from that spot.

October 2004: At a Halloween party in 2004 I asked a friend of mine, Chris Brethwaite, who is a paranormal investigator in the Kansas City area, if he could sense where a portal might be in the house. I told him nothing else. He found the spot within two minutes. Chris said he could feel a difference in the air and that it felt like a vacuum. He could not see it, but definitely knew where it was. I then asked another psychic friend of mine to come in the room and he located it in the exact same position. He also reported a vacuum like feeling when he placed his hand in the area.

October 2005: While my husband was out of town on his annual fishing trip with his dad I stayed home to run the business. On Sunday evening I took the trash out to the curb for the Monday morning pickup and when I returned to the deck on the side of the house there was a large black creature sitting in the middle of the deck. It had the head and upper body of a very large cat or small panther and the lower body and tail of a possum. I estimate that it was approximately 30" in length not counting the tail. I thought for a moment that this was an actual animal. It looked at me, then ran down the steps of the deck to the driveway and disappeared into thin air. I am sure this was a creature that entered from the portal.

September 23, 2006: I went out to the hot tub for my nightly spa (a necessity for people who work long hours) and almost immediately noticed a large 8" wide reptile-like eye, complete with detailed eyelid, pupil and vertical iris at the portal area. It simply looked at me for a while then disappeared. I closed my eyes for a minute, then opened them again and saw a bright white rod shaped object about 6-8" in length over the sliding glass doors to the right of the portal area. It stayed for one second, then darted into the wall over the doors. I got the distinct feeling that it was observing me. Obviously, the portal continues to open.

We have wood lattice work around our deck and I have noticed that while meditating I can see into another dimension by looking through the lattice. So far, I can only do this at night. I have seen other people and animals walking or moving about who seem oblivious to me. Sometimes I see eyes of different creatures looking in at me as if they are looking through a window. At times a whole face will be visible, and some are not human. I have asked my guides what these creatures are and found out that they are 5th and 6th dimensional beings that share the same location as we do but because of their high vibration we do not normally see them. They are simply observing me when I notice them, but normally they do not watch what people do on this plane. I catch a glimpse of these entities while in a meditative state, but not when I am going about my normal daily business.

Aliens or Ghosts? August 29, 2006: My husband went out of town on an annual fishing trip and I was home alone. I should note here that when unusual events occur he is not in the house.

One evening, I was awakened at 3:00 AM to a loud, clicking sound coming from the doorway in my bedroom. It sounded like it was approximately 6' up from the floor. I was immediately wide awake and knew that something was not right. The sound was a series of six clicks that sounded like an animal noise. Each click had a different tone and style.

This picture of a glowing orb by the fan blade was taken during a trance session to contact our deceased relatives.

I knew somehow that this was a communication from one type of creature to another, but don't know why I know that. I felt the presence of two beings in the room, and looked around but saw nothing. We have a skylight above our bed and the moonlight was adequate to see everything.

A few minutes later I heard someone walking on our gravel driveway, and then the hot tub jets came on. Next, I heard a crash in the kitchen sink. Oddly, I fell back asleep when normally I would have turned on the light and gone downstairs to investigate something like that.

The next day I checked the sink and nothing was there. The cats were outside so they couldn't have made any of the sounds in the house. I called my daughters and found that they had experienced some very odd things the same night. My youngest daughter, R. had a baby prematurely and he was still in the hospital. She had to pump milk every four hours to take to him and was up at 12:00 am in her living room with the machine on when she heard a loud banging on the wall near the front door (This is our rental house). It was a methodical banging and she thought someone must be playing a joke on her. R. ran upstairs to see if either of her boys was kicking the wall. Both boys were fast asleep and not moving. The noise stopped when she left the chair, but she called her older sister anyway.

R.'s sister and her husband raced over to her house and walked the perimeter, then drove around the block and saw nothing. They headed back home, but the banging started up again. This time, R. called the police and her sister, and Mia returned to R's house. Neither of them tried to call me but they wouldn't have been able to get me anyway because our phone was out due to a storm. The boys were still fast asleep. I did a reading on the house and could not determine the source of the banging noises— which is odd because I can usually pick up *something*.

Mia and her husband went home and went to sleep. She was awakened at 5:00 AM by a bright green light the size of a golf ball in front of and apparently coming from the satellite receiver. Normally, the green light on the front of the receiver is very tiny (1/4"), but it was very large and bright at this time, then it went back to normal size.

Mia noticed that she was wide awake, and normally she would be very groggy at that time of the morning. *This is similar to my feeling of being wide awake when I heard the sounds at 3:00 AM.* Mia then noticed a bright, blinding white light coming underneath her bedroom door, but when she got up to open the door there was nothing there and the white light was gone.

Why all three strange events occurred on the same evening I do not know, but it seems very bizarre.

September 15 25, 2006: Our phone goes out whenever it storms. Somehow water gets in the line and after a few days it dries out and comes back on. This time, however, the phone did not come on for over two weeks so I called the phone company, who said they would come out the next day. That evening, while I was in the house alone the phone rang twice with the rings being close together. Only one phone rang—the others did not. At first I thought that SBC must have come out early and fixed the line, but when I picked up the phone it was still dead. The next day, the technician came out and fixed it and said they had not been out before. I don't know why or how that phone rang, but it was during a T.V. show I was watching on ghost hunters. Was it a message?

The next day the phone was out again. I called Birch to ask if they could do a line check and they said that the phone was off because I ordered a disconnection. I did not call them and neither did my husband. Now I am trying an internet phone service and we'll see what transpires with that!

October 12, 2006: My husband fell asleep on the couch at 9:30 and I got in the hot tub outside on our deck. I felt as if something was watching me but all was quiet. My dog, Gracie, then started to jump on the wood gate to her kennel, frantically trying to get out. I looked over in that direction and saw a 9' tall human-like dark figure walk with a strange gate from one side of the shed to the other and through the wood fence into Gracie's pen. I know the height because I was able to measure a point on the shed the next day where the top of the head was located. The dog immediately stopped barking and trying to get out. I was concerned that something hurt her and tried to get my husband to go investigate, but he wouldn't wake up. So I locked the doors and went to bed, hoping that whatever it was would go away. This is the first time I've seen a very tall figure like that and I don't feel that it was human.

January, 2007: I got in the hot tub at around 10:30 pm and immediately felt like I was not alone, although everything was very quiet. There had been a recent snow and ice storm in the area, so it was quieter than usual. I then heard a clicking sound coming from a few feet behind me in the driveway, but since the deck has lattice and solid walls around it I couldn't see anything. There were a series of 10-11 clicks of different tones, similar to the clicking I heard in my bedroom several months before. It sounded like the noise was coming from approximately 6' off the ground. I felt that the sound was some sort of communication between two beings.

I practically froze with fear, which is unusual for me. I sensed danger, and immediately went in the house without looking in the drive. Normally, I would have stopped to look to see what it was, but I knew deep in my soul that the creature that made that noise was not anything of this world.

August, 2007: On August 18, 2007 I was sitting on the porch at around 10:00 p.m. when I noticed a flash of light in the sky just above the trees in our back yard. It looked similar to lightning, but was a very small, thin, white line. If I held something up to cover it at arm's length, it would have taken my little finger to do it. There were no storms in the area. Intrigued, I continued to watch the area for more lights, and sure enough, during the 20 minutes I stayed on the porch, I saw multiple flashes of tiny electrical charges of varying lengths and flash times. The longest flash time was a full two seconds. This seems to be longer than normal lightning. The flashes of light appeared around and up to approximately 15' over the trees but nowhere else.

On August 19, I returned to the porch in the late evening once again and noticed tiny white flashes of lights—this time over a tree not 20' from the house. It appeared that they were very close to me and not something far away. They also kept close to the tree, and did not appear anywhere else. I continue to observe this nightly.

January, 2008: I was sitting in the living room watching TV and glanced up to see a white apparition of a tall man walk in front of the living room door and head towards the laundry room. I jumped up to grab my Nikon DX 40 camera that was sitting on the dining room table nearby, and took some photographs. The camera would not work at first (oddly) but after I turned and took one picture of some flowers on the dining room table, I was able to take three more pictures in the direction of the laundry room. The camera was set to take three pictures at a time very quickly, so these were taken less than a second apart. The third photo (not shown) was completely whited out. See the photos on the next page.

October, 2009: I had a "Girls Night Out" party at our house and at one point three of us were standing in the kitchen when Mia reminded me about the voice that spoke out loud about the drawer shutting when she was younger. I reenacted the scene, not expecting any reply since we thought the ghost was long gone, but he once again verbalized in a deep voice "It was me." None of us were expecting that, and four people heard it. At Christmas I had the family over and we told the same story, and once again, our prankster ghost became audible and spoke loudly. This scared the entire family!

2010 – 2018: For the past eight years paranormal activity has not let up – in fact, it has increased. It seems that our house is a portal or gateway to anything and anyone who wants to pass through. We bought a house that is a paranormal hot spot! I've seen UFOs, ETs, shadow animals, unknown creatures, brightly lit rods and orbs, and dark webs in the house and outside the house.

After conducting numerous investigations in Independence, I'm convinced that the RLDS spiral temple has created a vortex of energy in the area and our house sits within ½ mile of this structure. I've received numerous reports from people within this ½ mile radius who are seeing many of the same things I do, and more. One person reported seeing two extraterrestrials looking in his window, and another reported seeing two monkey-like creatures near the temple.

A man and his two children said that they saw a huge UFO hovering over the temple one night and they watched it for about 15 minutes.

Another witness waited in his car one evening in the Temple parking lot when he saw a red ball of light come towards him and hover over the car for about 60 seconds, then move backwards towards the Temple building where it appeared to originate.

These bizarre stores are too much to ignore – I believe that the Temple is causing some sort of energy field or vortex that allows other-worldly creatures to come and go between dimensions.

This house continues to amaze me with new and perplexing anomalies. At the time this book is going to print we have had frequent ghostly encounters, heard disembodied voices, loud crashes, doors opening and closing, and most recently I saw the strangest thing – my cat started acting strange and staring at something on the wall that I could not see. She moved her head back and forth as if watching something move. I thought it was a moth and looked but nothing was visible. Right after, I saw a small 3" tall stick-figure ran across one of my side tables and jump into the air where he disappeared. I ran over to the spot where I saw the figure last but there was nothing there. Just when you think you've seen it all…

Photo taken just after a ghost sighting, with what appears to be a ghostly set of stairs that are not currently in the house.

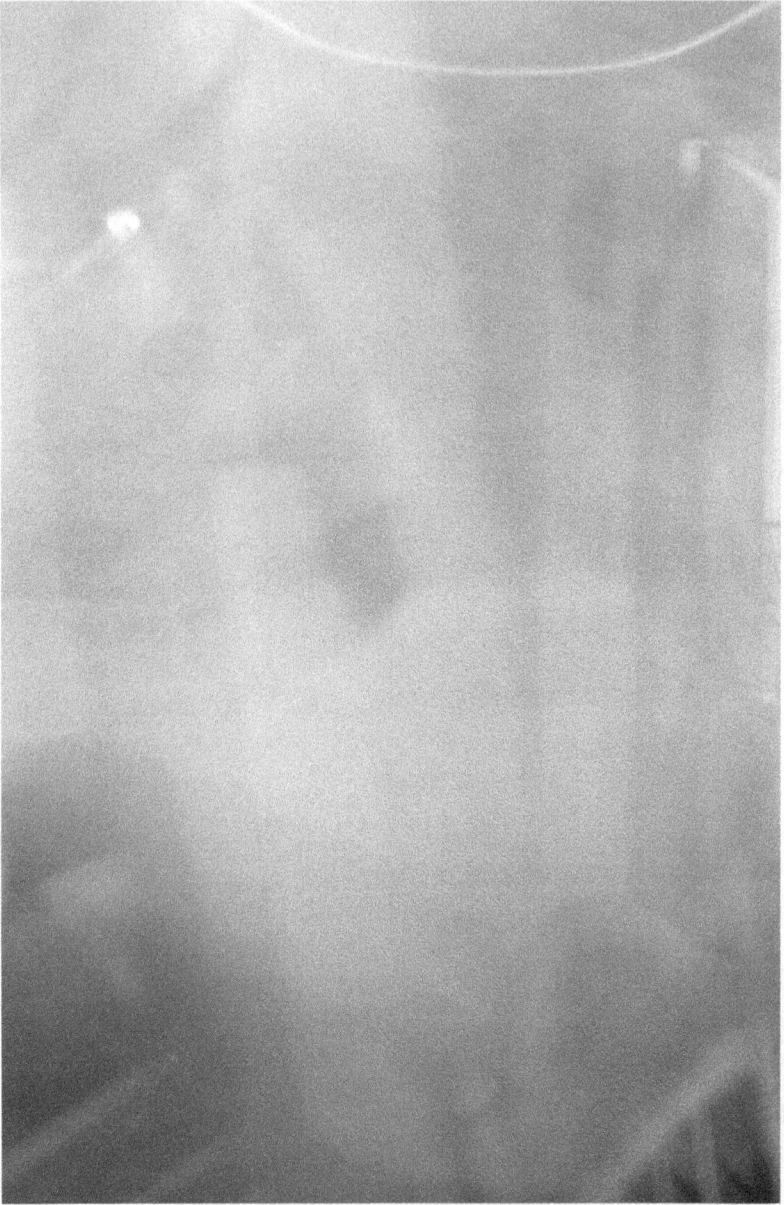

Second picture taken from the same angle, with what appears to be a distorted hand in front of the camera lens, as if something was trying to stop me from taking a picture.

The Pope Street Demon
Former private residence

912 S Pope Street (now razed) was one of the most haunted locations I have experienced. The house was at least 100 years old at the time of our visit. My oldest daughter, Mia, rented this home in March of 1995 for a period of one year. She lived there with her husband and two small children. Mia was pregnant at the time and had her second child while living there in January of 1996.

Soon after moving in, Mia asked me to visit the house because she felt that it was haunted. She had already seen several apparitions and had poltergeist activity in the kitchen area. A black ghost cat roamed the bedroom, leaping from the dresser to the bed, and then running through the wall. There was a water leak that started suddenly and she could not get it turned off. She also saw the window in her room disappear and turn into a door. Several entities walked into the house from this location.

A few minutes after my arrival to investigate in March of 1995 several dishes fell to the floor, even though no one was nearby. I felt the presence of several angry ghosts, so I went into trance to see who they were and what they wanted.

The first spirit was that of an older man named Peter who built the house. He felt much attached to it and felt that anyone living there was an intruder. I asked him if he knew that he was dead and he said yes but it did not matter, this was HIS house. Peter was the one making noises in the walls, turning water off and on in the bathroom and kitchen, and breaking dishes. I asked him if he was ready to leave and go to the light. He wanted to think about it and said he would talk to me if I came back later. I knew he was lying about wanting to talk to me again.

There were several other spirits at the location that are much older and have been there for hundreds of years—way before any houses were built in the area. They are Native Americans and are protecting the location.

I then turned my attention to something else that caught my eye while I was speaking to Peter. I saw several small "people" that looked more like alien grays. They moved back when they saw that I could notice them. It seemed as though they were very surprised that anyone could see them. I could sense that they were there to observe my daughter and when I asked about it they said yes, but it was none of my business, and to leave them alone because they had a job to do and they were not harming anyone.

I left to ponder what to do about the situation. The next day, the water pipe under the sink burst open, flooding the house. There was not shut-off valve for the water in the house. When the city came to turn the water off at the street the valve was rusted shut and they could not get it closed for several hours. As a result, the entire house was flooded.

After discussing this with my daughter, we used some techniques to protect her and her children. It worked, and she did not have any visits from the "aliens" or ghosts again, but still felt their presence. She moved out a few months later, and heard from the landlord that he could not rent the place to anyone else because things kept happening to the house that needed major repair. He finally gave up and tore the house down completely.

I drove past the location where the house was a few months ago, and still feel negative energy there. I think this is a portal area where entities can come and go at will, no matter if there is a house there or not. The owner has fenced the property and put up "No Trespassing" signs. I wonder why the concern for a vacant lot with nothing on it?

Evil at Sterling House
Private Residence in Independence
Extremely haunted

The owners of this property did not know what they were getting into when they purchased it in 1996. Since the family of five moved in to this 1929 house near 23rd and Sterling they have been increasingly harassed by resident spirits.

It started with occasional 1950's era music coming from the basement. No source for the music can be located, and there are no radios or TV's in this area.

One evening their three-year-old boy became very upset when his mother walked into the kitchen and "made his daddy go away." His father was at the store and there was no one in the house. The child insisted that he was talking to his father.

The young children of the owners often speak to and play with several ghost kids, and give their names and descriptions. Sometimes they even argue with the ghost children, or tell them to go away, as normal kids would act.

Their five-year-old recently made up new names for the ghostly children in an attempt to appear to be talking to imaginary friends, rather than real friends. The new names are popular cartoon characters. The wily child did this in an attempt to keep her other family members from being concerned about her interaction with the ghosts. The parents believe that the ghost children may have suggested this strategy to their young daughter.

The owners have found handprints on the foggy mirror in the bathroom and one on the glass of the back door. The bathroom water faucet is often found on in the middle of the night.

Their 14-year-old daughter suddenly woke up at three O'clock in the morning one day to see a shadow person standing in the middle of her bedroom. The shadow was watching her TV that had been turned off before she went to bed. The girl stayed very still and made no noise, but apparently the shadow person knew she was awake anyway. He turned and looked straight at her, then slowly disappeared. She described him as having no features, being all black, and having mitten-like hands.

Recently, dark shapes of adult entities have been interacting with the owners and one even attempted to strangle the wife with a scarf one day, which she felt go all the way through her neck before the entity disappeared. She had been physically pulled across the bed by her neck, and there was a red mark on her neck. She described the being has being *all black and wearing a hat and long frock coat.

During several visits to the home, our QUEST team has detected high EMF Readings and changes in the ambient temperature that cannot be explained. We have seen unexplained shadows, heard footsteps on the stairs, heard the 1950's music, and even witnessed a door slam. I did a clearing of the home which resulted in a cessation of paranormal activity for about 12 months but recently the activity has picked up again. The owners have begun seeking another home to move into.

*Note: See more information on shadow people in my book "Gateway to the Dead: A Ghost Hunter's Field Guide."

Crysler and Lexington Ghosts

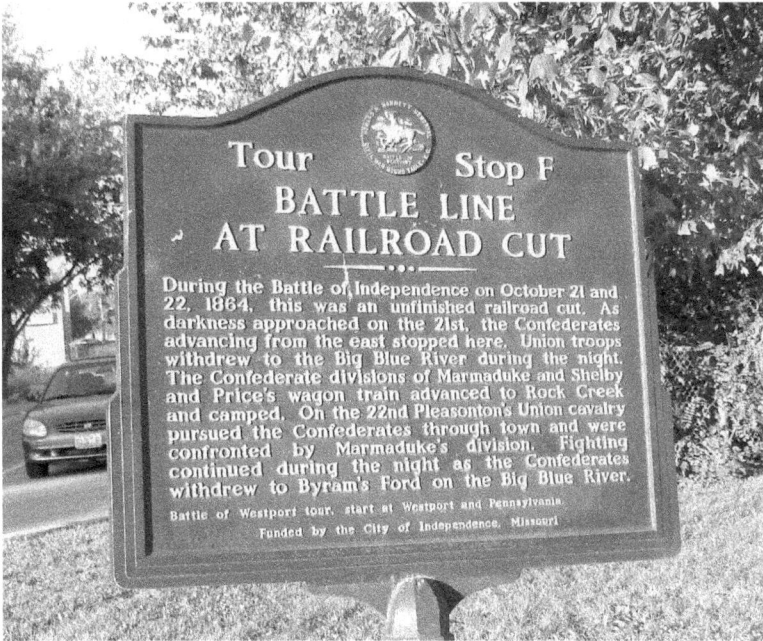

There is a 100–year old+ house at the corner of Crysler and Lexington that was made into three apartments. There are two apartments on the lower level and one unit above them. This house is located on the exact spot where a civil war battle was fought near a railroad cut in 1864. I was called to investigate paranormal activity in both lower level apartments in 1996.

I started my first investigation of this property on the south side apartment of the building. The tenant, C.T. told me that she saw a dark shadow in the shape of a man move across the living room/dining room area on several occasions during the day and night. He did not appear to notice her. She reported that she also heard scratching and banging in the wall behind her bed and in her bedroom closet. Her two small children refused to go near the closet, and she had to get out their toys and coats for them when they needed them. C.T. said that the activity increased in frequency since she moved in 12 months earlier and that it felt more menacing as time went by.

Immediately upon entering the apartment I sensed a heavy presence that seemed to move around the room. I went into a trance and saw a man wearing a long black trench coat, pants and boots that looked like 1860's attire standing in the dining room area. He had an angry look on his face and was looking right at C.T. The man said he did not want her or anyone else in this location because it belonged to him. He walked to the closet area. I could see that he didn't live in this house as it was, but lived at that location in a small cabin many years ago. The closet area was likely the original site of the entrance to the log cabin and was where he walked in and out of this world. I did a clearing and afterwards, the spirit was quiet but he remained there.

I then visited the neighboring apartment on the north side to see C.C. who told me that she was experiencing banging noises in the wall adjacent to the right-side apartment and heard loud footsteps in the vacant apartment above her. When she looked inside the upper level apartment there was no one there.   She also saw her kitchen cabinet doors open and close on their own and saw a white shadow and a dark shadow on several occasions. C.C. said that the kitchen cabinets flew open and slammed closed, and dishes flew out of the cabinets and came crashing to the floor the night before my arrival, which is what finally prompted her to call me.

After going into trance I saw the same man from the right-side apartment passing through, and also a petite woman. The woman had no relation to the male entity and had died recently. She told me that she was trying to get C.C.'s attention because C.C. is psychic. She wanted help of some kind because she felt wronged. I was able to see what she needed help with, but we were unable to assist her.

I visited the apartment at a later date to do a clearing, which was unsuccessful. This happens when an entity is very strong-willed and thinks they have unresolved issues that must be taken care of before they move on. In this case, her spirit is unwilling to leave until things are resolved to their satisfaction.

Both tenants moved out after their 12-month lease was up due to ongoing paranormal activities at their apartments.

During a recent visit to the property I went in to a trance state and saw several civil war encampments in the area with men in confederate uniforms. There were teepees of guns next to tents and burning fires with cooking utensils all around. I saw the men fire their guns, and noted that I could see flame come from the end of the guns when they were fired. The noise was deafening. I could see men on both sides get injured and fall. I saw injured men being carried away on stretchers and loaded into horse-drawn carts. They were headed to a nearby hospital, which was a large home that had been converted.

The soldiers fired across the railroad cut, which was unfinished at the time. There was intense and continuous fighting and many spirits still roam the area. It was extremely difficult to see the pain and suffering these men endured and I had to leave after a short time.

There is a place to stop next to this location on the West side of the corner where the marker is located. If you like, step out of the car and walk around the area and see if you can feel or see anything. I think this would be an excellent place to ghost hunt because of the extreme emotion that is left behind after a battle.

Paranormal Events on East Lexington

This is truly one of the strangest cases I have ever worked on, and I think you'll find it extremely odd, too. In late summer of 2003 I received a call from an acquaintance, who was very panicked. She is a housewife, and lives with her husband and two dogs. In her 63 years she had never seen anything like this.

Sara D. was awakened at approximately 5:30 a.m. on a Saturday morning by her dog, which needed to go outside. It was still dark outside and it was unusual for the dog to wake up that early. Sara got up and let the dog out the front door, then waited for him on the front stoop. She looked across the street and noticed that there was a newer black four-door sedan parked in front of her neighbor's house two doors down and across the street. She thought it was odd that the parking lights were on. Sarah noticed a male driver and possibly a male passenger in the front seat of the car, both wearing dark clothes. The passenger turned around and looked directly at her, and Sara could see that he was wearing sunglasses.

Sara then looked at the house, expecting a person to come out of the door to the waiting car. Instead, she saw a small, 6" - 10" diameter bright white light pass through the closed door of the house, then travel to the waiting car and enter through the back passenger side window. Immediately, Sara saw a third person sitting in the back seat who wasn't there before. The car took off quickly. Sara said she could swear that the light became a person when it went into the car. She thought that the quick retreat from the house was a reaction to her watching the car.

I went to the location the next day and "scanned" the area to see what I could pick up. I did not feel the presence of any spirits in the house or see anything unusual at the time of my visit. I did, however, sense that the people in the house Sara mentioned were being watched, but by whom (or what) I could not tell. The black sedan, men wearing black suits and black sunglasses certainly fit the description of "Men in Black." I do think that Sara is telling the truth about what she saw. She and her husband have since moved away to another state and oddly, their house burned down while it was vacant. Whoever or whatever burned the house has not been found.

The Shadow People

On the 600 block of East Lexington is a house haunted by unfriendly spirits and shadow people. The family told me of events dating as far back as 1976 when J.D. and his wife, Bettie, moved in the house with their two young boys, J.R. and A.

Not long after moving in, the entire family noticed that they felt like they were being watched all the time. None of them could explain it. While the husband and boys were away at work and school during the day, Bettie would notice this feeling more often. After a few months, Bettie noticed that things would get misplaced, only to turn up in an odd place. Keys would be missing from the key hooks by the door and turn up underneath a laundry pile. Her hair dryer disappeared one day, only to show up on the back porch a few days later. Her husband said that she must have moved these items and forgot about it but Bettie knew better.

In 1977, the youngest boy, A., age 10 started complaining that he was now afraid of the dark and would like a nightlight. This was odd since he had not been afraid of the dark before. He said he saw red eyes watching him at night. At first, the older brother, J., did not believe him but he soon started seeing the red eyes as well and also wanted the nightlight on. Soon that was not enough, and they started leaving on the table lamp.

The boys started losing sleep because of the "monsters" in their room.

Their father still did not believe that anything unusual was going on and he did not experience anything -it seemed that the spirits were not interested in showing themselves to him.

Approximately two years after moving in the house, Bettie saw a Cheshire cat in the tree in their back yard. She thought it was her imagination at first and tried to ignore it. The cat looked like a cartoon cat with a big smile. It showed up on several occasions when she was alone, but would never appear when any one else was around. Bettie was afraid to hang her laundry outside anymore and asked her husband to purchase a dryer. He refused. She did not mention the cat for fear of sounding crazy.

As the years went by, more and more activity increased. Finally, all of the family members saw the red, glowing eyes and dark shadows in the upper bedroom that the boys shared. They all experienced hearing footsteps walk across the upper floor when everyone was downstairs. And everyone experienced missing items that would appear weeks or months later in odd places. An overwhelming feeling of negative energy in the house pushed the two boys to leave home early because they could not stand to be in the house. As soon as they each turned 18, they moved out. The parents opted to stay and hoped things would return to normal. They didn't.

Months would go by with no activity but as soon as anyone noticed that things were quiet around the house, something new would happen. It was as if someone or something was listening and reacting to the conversations of the homeowners.

Our paranormal research team was called to investigate the house in 1999. We arrived at the house, but none of us could go inside except Mia. We all felt an overwhelming heavy, dark, presence. I decided that we would do our investigation from another location and do remote viewing. Remote viewing is a term for observing things from close or far away, whether it is a city block or around the world. This is a technique that I used often in my investigations to find missing persons. While none of the rest of our team had used the technique before, they all quickly learned how to do it.

We split up and I asked each team member to see what they could find out on their own. When we compared notes, it was surprising even to us how similar our findings were. Each of us saw a dark entity that could materialize at will. None of us felt it was human, but that it was likely a Shadow Person.

Shadow People are non-human entities that observe people and take energy from them. They can only live on fear generated by people, hence the menacing red eyes that would certainly scare anyone, especially a child. We all felt that the other paranormal activity in the house was generated by this dark presence that was growing in strength, and also by several spirits of people who had passed on. I sensed that some of these were Native American spirits and they were angry about their land being disturbed.

I could only advise the family to leave the house as soon as possible as I had no luck getting rid of this type of Shadow entity, and research into the topic was fruitless, with other investigators warning everyone to stay away from shadow people.

I have observed two other Shadow People at other times, the first being in 1978. It was in the shape of a man who stepped out of some trees while my husband and I drove through the neighborhood. The man walked to the center of the road, and then disappeared in front of our eyes.

The second incident was in May of 2006. I saw a man wearing an old 1900's western-style frock coat and low wide-brimmed hat standing on the steps of the First Presbyterian Church on Lexington. The man was staring angrily at the RLDS Temple site. He was all black from head to toe, including his skin and clothing. The man stood there for a minute while I was stopped at a stop sign, then disappeared. I was quite shaken up by this incident and it still bothers me when I pass by this church. I had the strong feeling that the shadow person was Joseph Smith, III.

After visiting Mount Grove Cemetery where Joseph Smith III is buried, I feel that his ghost does haunt the place occasionally due to some unfinished business. Attempts to contact him have proved futile because he is very angry and is not cooperative.

The Strangest House in Independence

Private Residence in Fairmount

Fairmount is one of the oldest neighborhoods in Independence, and was once the hangout of Frank James and others at a general store nearby. It is located off of 24 Highway. This area was a pioneer village prior to being divided up into Plats in 1860. This particular house was built in 1910 on top of a much older rock foundation and basement. The homeowner's dog digs in the yard and has retrieved farm implements from the early 1800's. The owners have asked that their address not be printed in the book and do not want visitors.

I was first called to investigate this house in 1987 and have returned many times since then. The activity continues to this day. The homeowners, F. K. and his wife, A., have experienced spirits materializing on numerous occasions. I have seen one spirit of a man in a long old-west frock coat wearing two pistols. He has long hair and a beard. The man walks up and down the creaking stairs of the house, and can be heard at all times of the day and night. Since the current house was built after the time period of the clothing this ghost wears, I assume that there were steps in the prior house in the same location. This spirit means no harm, and seems unaware of anyone else. He is very serious about his work, and seems to be concentrating hard while he walks. I believe he was a deputy or marshal.

One night shortly after retiring for the evening, A. told. F. to look at their dresser. There, in full Confederate Civil War uniforms and swords were three tiny men standing on the dresser. None of the men were more than 24" high, but all were in perfect proportion. These three ghosts have appeared several times and they do react to the homeowner's when they notice them. A. and F. do not feel any malice from the spirits and do not want them to go. This was my first investigation involving ghosts appearing in miniature, and I honestly do not have an explanation for it. Obviously, the current house would not have existed during the civil war, but the foundation is very old and may have had another structure on it previously.

In May of 2006, F. came into possession of his grandfather's white suit from the 1940's. He hung the suit on the outside of the closet and went to bed. A few minutes later, A. saw something in the suit and told her husband to look at the suit because his grandfather was standing in it. A. had never seen F's Grandfather, so he asked her to describe him. She did and it was exactly correct.

F. could not see his grandfather, but felt his presence. A short while later F. felt someone lie next to him and put their arm around him. At first, F. thought it was his wife, but when he looked, she was too far away. He sensed that it was no one he knew and this presence had nothing to do with his grandfather, but was a different spirit. F. slowly reached around, grabbed the hand and moved it off of him. Frank said the hand was very pliable and felt something like jelly. He was too afraid to turn around to look at what it was. He is thankful that he has not experienced this since.

Trance work in 2007: I did a scan during a trance and saw a lot of activity over the years. There have been many people living on this spot almost continuously since the 1820's. I saw tents, then a cabin, then saw the cabin being torn down because it was not built very well. Later I saw a foundation being dug and a house built on top of it. That house burned down and was replaced by the current house. At one point, the property was seized (stolen) by someone who kicked out the previous rightful owners.

Trance work in 2008: I got a clearer picture of the area this time and was led to the back yard to an area just beyond the cement patio. Some of the spirits said to tell F. to dig up an area there and he would find some very old stone steps leading downward. I asked what was at the bottom of the steps and I saw wood—the top of a coffin. I opened the lid and there were remains inside. Then I saw wood coffins all around the back yard and the neighbor's back yard directly behind this property and a big house east of this area. I was looking at the area when it was a larger piece of property and before it was split up into plats. F's house was built right on top of an old family graveyard!

Historical research: F. did some research on at the Historical Society and found that the area was a favorite of Frank James' and that a store he frequented used to be located only 2 blocks from this house. It was a hubbub of activity, so it is no wonder that there are so many spirits here.

5/3/2007: The owners heard the sound of a music box coming from the upstairs bedroom. They do not own a music box. They also heard footsteps walking across the carpeting and stairs for several days in a row. The cat and dog acted strangely during the same time period—often following an unseen object around the room.

05/21/07: Three of the QUEST Team members re-visited the house at the request of the homeowner, who reported recent activity. The team stayed for over three hours, during which time the ghosts were very active. I first sensed something by the fireplace, where I then tested the area with the EMF meter. The meter spiked at the opening of the fireplace.

The three of us felt activity upstairs, so we all moved to the upper floor. As we walked up the steps, our team member Rachel felt a presence in front of her. I should note here that I have seen the spirit of a man on these steps on several occasions. Rachel used the EMF meter, and it spiked, and then stopped. As she moved forward, it spiked again. She was able to follow the spirit down the hall and into the far bedroom by using the meter. We had not experienced this before, but this was a very strong spirit. Upon entering the bedroom at the end of the hall, I saw it as it used to be approximately 100 years ago. There was a pot belly stove in the center of the back wall in front of an old chimney, flowered wallpaper on the walls, a simple iron bed, and one dresser. The man who lived there was tall and wore a long coat and pistols. I believe this is possibly a deputy marshal or sheriff who rented the room from the owners. I sensed that he reported to work each day in downtown Independence.

Mia, our third team member, then opened the closet door and picked up psychically that someone was once restrained and kept in this room. There was an ominous feeling about it. We all wanted to leave the room immediately. I do not feel this was related to the (possible) Marshal.

Investigative trance work: Next we moved to F. and A.'s bedroom. I had the sudden feeling that I should lie down on the bed and close my eyes, and did so. I fell into a half-trance and saw a single mother living in the house with an adult daughter. The daughter and mother were arguing over the daughter going somewhere by herself with no escort. There was no father in the house—I sensed that he was dead. The women both wore long flowered dresses with petticoats and bustles that could have been 1870's-1890's attire. Why this was significant, I don't know yet.

We all moved into the hallway and as I glanced back into the bedroom I saw a column of whitish/green fog appear. I pointed it out to Mia and Rachel and the homeowner, who all saw it, too. It dissipated within a minute. We then looked down the hall to the far bedroom and saw fog again.

F. asked us to look in the bathroom where he sometimes saw faces in the carpeting, but none of us saw anything. However, I felt the presence of several spirits.

We then moved downstairs and sat at the dining room table, where a column of fog suddenly appeared next to the table. I took a photo of where the fog appeared, which was to the right and in front of Mia, who was standing at the time. Next, we saw the fog rise towards the ceiling and disappear, but I took a picture anyway. When it was printed out, there was an unexplained rod-like light on the ceiling (see the next page), which was not visible at the time the photo was taken.

A rod appeared in the dining room

May 31, 2007: F. reported that he was putting laundry away in his bedroom in the late afternoon when he heard the name "Zizzi" called out by his grandfather's voice. F. had been trying to contact his long-dead grandfather through meditation for some years with no success—until this occurred. He heard no other words. "Zizzi" means "aunt" in Italian, and is what his grandmother's family called her. F. is absolutely certain that it was his grandfather's voice.

June 1, 2007: F. reported that as he and his wife went to bed and that both noticed a black, triangular shaped object floating in the room. It appeared to be approximately 12" across and according to the homeowner was shaped "kind of like the stealth bomber." The object darted about the room, hesitating over several areas, with most of the concentration being over them as they lay in bed. It appeared to be taking pictures since a bright flash of light came from the object several times. Then the object moved through a closet and into the chimney. The episode lasted approximately 10 minutes and the homeowner's said they felt as it if was "scanning" them and the room.

June 2, 2007: F. reported that a small triangular shaped object appeared to them again, only this time it had a greenish glow about it and it stayed for a few minutes longer than the previous evening. This one also moved about the room, then darted through the closet and into the chimney. *Note: See the chapter on haunted chimneys.*

October, 2007: A. told me that she sees ghostly nightgowns and heads of spirits floating in her bedroom at night. F. and A. both saw the full apparition of an old lady in old-fashioned (early 1800's clothing and bonnet) who turned to look at them, then faded away. *Note: this apparition has since appeared on numerous occasions.*

October, 2008: F. said that the strange activity continues, and that he has seen several small UFO's in the house with different features. They only appear in the bedroom and only after his wife has fallen asleep. Some of the craft are saucer shaped, and some have wings and look similar to a plane. I noted the strange correlation with the small Civil war men that appeared to his wife, - and the small UFO's, which are both extremely rare types of sightings.

I decided it was time to find out exactly why there are so many spirits here and asked the couple to leave the house for an hour one evening. As I got comfortable on the couch and started to do trance work a black, shapeless shadow moved across the ceiling of the entryway. I knew I was not alone at that point. I meditated and contacted several of the spirits who told me that there is a portal area near the chimney in the bedroom here, and that most of them are ex-residents of other houses in the neighborhood and of the old family graveyard that is buried in the back yard of this house and the house behind it. The spirits use this portal to move in and out of their dimension and sometimes congregate together here. That explains why the occupants see so many non-resident ghosts at this location. They are simply using this spot as a gathering place and to use the portal travel to our dimension. Most of the spirits are not "haunting" the occupants, they just happen to be seen sometimes as they pass through.

I encouraged the owner to dig in some spots in the backyard to see if there are indeed graves there, but so far, he has not done so.

2012: The owner took a photo of her yard one day when she saw an unexplained foggy mist appear over some of the flowers. There was no fog anywhere else.

2018 Update: Paranormal activity continues at this house. The owners still see the "regular apparitions" appear and disappear, but they also have seen their deceased cat and dog appear on numerous occasions. At times the animals appear transparent, other times solid. They leave indentations in the bed, and both owners can feel the animals at their feet. One evening recently, their deceased dog appeared at the bottom of the bed where he used to jump onto, then the dog walked on top of A. and curled up on her neck and shoulder just like he used to do when he was living. This time, the dog was fully formed and felt solid, and A. was able to pet him!

At this point, it appears that the area is another portal which allows entities from other dimensions to pass through easily.

The Haunted Office

1136 S Pearl Street—not open to the public

This house, built in 1910, was converted to a zone C commercial property and beauty shop in 1970. We purchased it in 2000 and made it into an office for our staff with the intention of eventually putting up a larger building and tearing down the old house. We moved in and almost immediately knew we were not alone.

The first incident occurred to my secretary. She was missing her scissors, tape dispenser, and stapler. We looked everywhere but could not find them. The next morning, they were back in their proper spots. This activity continued for three years, with desk items sometimes turning up in odd places like inside file cabinets.

Next, we noticed lights would go off and on of their own accord. And finally, a white mist began to appear to my assistant, then to the rest of us. One day, while she was alone late in the office, my assistant, Tamie, said she saw an old lady walk by her desk and go into my office. The lady was transparent but clearly visible.

I began to see the lady not long after. My night bookkeeper reported seeing someone walk through the front office and go into my office on a number of occasions. One evening she took a digital camera and took some photos of my office. In one photo, there is a white mist clearly visible, and in another there are no pictures on the walls, but they are there normally.

We noted that every time we spoke of tearing down the building in the future to make this area a parking lot, it angered the spirit and she would react by turning lights off or taking things and hiding them. I have still not retrieved many items and wonder where they are.

I saw the ghost of the old woman walk through my office twice. She was wearing a dressing gown and holding a towel. I stood back out of the room the second time I saw her, and looked at the room psychically. This used to be her bedroom and she was walking towards the bath.

We did an investigation of the house, and found out that the old woman used to live here by herself for many years and was very attached to it. She died in the house. Apparently, she does not want us to destroy it.

During a trance, I saw other activity from long ago with someone else who owned the property during the 1800's. I have seen a man wearing a long frock coat and hat walking around the area surveying it before any buildings were erected. He just stood there and looked around.

We moved out of the building in 2005 and now use it as a temporary warehouse and mini-showroom for woodstoves. I drive by at night sometimes to find all of the lights on, even though we know they were turned them off. *Note: this house is next door to Pearl House #2 where a ghost likes to turn lights on as well. He might be visiting us at both places.*

We have an alarm system and movement sensors on the building. In August of 2006 I was called by our alarm company, Barnhart Security, who said they had movement in the Southwest side of the building *interior* but no open doors or windows. I asked how that was possible, and they said they had no explanation since the movement sensor is set at 40lbs. (it takes something at least 40lbs. to set it off). They could not explain how there could be movement inside the building without any doors or windows being opened to let someone in.

My husband and I rushed to the building a few minutes later and found the place locked up securely. However, a ceiling tile had been moved out of place in the Southwest room and the lights turned on in that room only. The tile was not like that earlier in the day and the lights had been turned off before locking up in the evening. This tile is located directly below some boxes, a baby bassinet, and other items in the attic that we cannot reach, and frankly, don't know how anyone got them up there. Was the ghost looking for something or telling us to look at these items?

As of the time of printing of this book the lights in the building continue to go off and on of their own accord. You may drive by in the evening to see if you see any ghostly activity outside. It is now vacant but we plan to make it a meeting space.

Gladys' Ghosts

Private Residence

Gladys Ward lives in Independence. She did not want me to print her address for obvious reasons, although she has since moved out of the house. Gladys asked me to do a remote scan on this house after she experienced some poltergeist activity in the home, and some foggy shapes showed up in a photo she took of her grandson on his birthday in September of 2002. None of the other photos taken at the same time showed anything abnormal. She said she felt the house was haunted and had experienced many strange happenings while there so decided to sell it and get out.

A photo taken of Gladys' grandchild with three unknown misty objects in the foreground. Pictures taken before and after have no anomalous objects in them.

Gladys and her husband experienced poltergeist phenomena while living in this house which included doors slamming, loud steps, lights going off and on, and they could hear someone talking in another room but when they investigated there was no one there. Both also saw dark shadows moving about the home. They felt an eerie, creepy presence, especially at night, and both found it difficult to sleep. One evening, something pulled the covers off of the couple while they were just falling asleep. The couple knew there were ghosts here, but didn't know who they were or why they were being haunted.

One day, during remodeling of the basement, they removed the old wood paneling off of the walls that had been previously salvaged from a nearby abandoned church by the previous owners. Gladys found out from neighbors that there had been satanic rituals performed in the church, and they felt that some negative energy or spirits must have stayed with the wood paneling and was the source of their strange experiences.

It got so bad that the couple decided they had to move out because they weren't getting any sleep at night and were afraid to invite their families over any more.

I went into a semi-trance state and remote-viewed the house from another location because the homeowners did not want to have anyone else visit. Three spirits made themselves known to me—two aggressive men and a woman. I could sense a rivalry between the men and fighting between them with unresolved issues. The woman felt trapped and wanted to go on to the light, but was influenced by the men to stay with there with them. I believe that one of the men was injured in the head area but could not tell if it he died as a result of the injury. The three spirits seemed to be very connected to the space and did not want to leave.

Not long after Gladys and her husband finally sold the house, and moved to a newer home which so far has no evidence of any ghostly presence. "Which is exactly how I like it," says Gladys. "I've had enough haunting to last me the rest of my life." I sincerely doubt if they disclosed the information about the haunted house to the new homeowners, which by the way, is legal in Missouri.

I believe that these are the three negative spirits that linger in the house.

A Poltergeist in Sugar Creek
Private Residence

As true ghost stories go, this one is one of the scariest. J. and T. Garcia own this 90-year old two-story home with a finished basement. When the three grandchildren spend the night, the oldest boy, now age five, sleeps in a bedroom in the basement. When the boy was two he began speaking about a nice old lady who sat in the rocking chair in his room and visited with him. He kept this up over the years. From his description, the Garcías knew that this was J's mother, and felt glad that she was around. They thought she was there to protect their grandson from the dark energy they believed was in the house.

From the time they moved into the house over 30 years ago, the Garcia's have been occasionally harassed by a dark male figure that walks through the house and walls. The entity has been known to stop and turn to look directly at them in a menacing manner. At times his features are clearly visible, but at other times they are cloudy.

When they first moved in they saw this entity a few times, but it did nothing except walk around. After a few years, however, he began hiding small belongings, such as jewelry and keys, which they would find in strange locations such as the attic or garage.

When the grandchildren came along and started spending the night, the dark entity increased his activity in an apparent attempt to get them to leave the house. The Garcia's awoke several nights to find the dark figure standing at the foot of their bed. They began hearing crashing sounds throughout the house, but when they investigated could find nothing out of place.

At the time of my investigation of this site, the Garcia's had recently experienced objects being thrown at them and they decided that it was time to call in some help. First, I did a remote view or scan of their house while I was at my home, which is something I always do before going into an unknown situation. What I saw was the neighborhood at a previous time period, before any settlers reached this area. There were many Native Americans here, and many teepees scattered around, all not far from the Missouri River. I knew then that I was dealing with a very powerful spirit. I wanted some additional backup but none of the QUEST team members were available at the time.

Upon arrival at the house, I saw the ghost of the grandmother and spoke to her. She was indeed on the site to protect her grandchildren, especially the child who slept in the basement, because that was the area where the dark entity resided. Next, I smudged the entire house with burning sage to help remove negative energy, doing this because I knew that the entity would be familiar with this Native American method. That act brought him out and we spoke for several minutes. The spirit said that he used to live in this exact location. I was able to get him to understand that he had died many years earlier, and that his entire tribe had long since been gone. I explained that settlers built up this area and now had many houses here and just wanted to live in peace. He finally agreed to leave and go with his spirit guides. The entity has not returned since, and all strange activity in the house has ceased. While the grandmother's spirit still visits on occasion or leaves the rocking chair moving as a sign she is there, her visits are less frequent than before.

2018 Update: The family recently moved out and the children are now much older, but they still have occasional visits from their grandmother at their new house.

Old Man Walking
Private Residence

Jamyi T., one of our team members, lives in Independence near Sugar Creek, but does not want us to publish her location. She experiences hearing voices and seeing movement in her house at night when it is quiet. One evening in late 2005, she was talking on her cell phone with a friend while lying on her bed in the dark. She saw the ghostly apparition of an elderly man walk out of the bathroom, go around the corner, then walk into the wall where the side of the bathroom is located.

Thinking fast, Jaymi said out loud "I'm not afraid of you—please do that again." To which the ghost complied by walking the same path as before, only this time he glanced her way while she snapped a photo with her cell phone. The picture is not high quality but some people see the outline of a face or white fog. Jaymi said that in person this was very clear and that the apparition glowed somewhat. The ghost kept doing the same thing on a regular, almost weekly basis, and she became used to it.

In August of 2006 a friend was visiting Jamyi when the ghost again made his trek out of the bathroom door and around the corner. The friend witnessed it and immediately left, never to return. Jamyi was able to snap another photo of the spirit. Normally, I believe that this type of routine behavior is a Time imprint, meaning that it is something you see that occurred in the past however, with the spirit's obvious interaction with Jamyi by looking directly at her, I believe that this is a spirit that is attached to the house and is comfortable carrying out his daily routine. He must be haunting the location.

I sked the spirit if he knew that he was dead and he said yes, he did know he had passed on. He lived and died in this house and wanted to stay because he was most comfortable there. I told him that as long as he did not scare Jamyi I would not ask him to leave. So far, he has complied.

Native American Burial Ground

Susan Waters, owner of an Independence business on the Square, used to live in a house in the 1200 block of south Spring Street, just south of 23rd street and West of Noland Road. Susan, her husband, and friends, had numerous encounters with spirits while living in this home.

Susan had most of the experiences while her husband was away. One day she opened her closet to find her robe standing up right over her boots, as if someone was in the clothing. The robe was not on a hanger and was floating in front of the hanging clothes. She grabbed a broom and poked it, only to find no one there. "It was just filled with air." Susan said.

While a friend and her child were visiting on another day, the child played with a toy that made the sound of drums. She kept pushing the button over and over again to make the drum sound. After the toy was turned off, the pipes in the basement started banging to the exact same beat the drums on the toy did, and kept it up for some time. Her friend heard it, too, but when she called her husband in from outside the banging suddenly stopped. Susan was frustrated that her husband never had these experiences. But that would soon change.

One evening the homeowners went to bed, and soon after heard Susan's guitar strum. The guitar had been placed in a laundry basket in their room with towels in it. The owners turned on the lights to see what caused the strumming sound, and could find nothing. They went back to bed, and Susan's husband asked her what she thought caused the guitar to make that sound. Her answer was "It must be the ghost in the house." Right after she said that the guitar strummed yet again, as if confirming her statement.

The Waters did some investigating and found out from an older neighbor that grew up in the area that the entire neighborhood once was a

Native American burial ground. Susan and her husband have since moved to another house where they have had no experience with ghosts.

Our Haunted Rental House

This house is one of our rental properties but is currently used as our office. We purchased it in 2003 to use as a showroom, and then turned it into a rental. The first family who rented the house stayed for the contract period, but left after a year due to paranormal activity.

I saw a man on the roof one evening and thought it was a prankster, until he disappeared right in front of my eyes. On at least six other occasions I saw a man looking out of the dining room window, but no one was in the house when we investigated. My employees saw someone looking out of the windows, too. Lights seemed to go on or off at night of their own accord, and we all heard footsteps on the basement stairs. My secretary said she saw a black cat at the top of the stairs one day, but there was no cat in the house.

I decided to do an investigation and called the team out one evening. Three of us went into trance and got similar information. During our session we all heard loud footsteps on the basement stairs, but when we went to look there was no one in the area. We searched the whole house but found nothing. We did pick up some muffled voices on the digital audio recorder.

All three of us got information psychically that there was an older man by the name of George who lived in the house alone for many years after his wife died. He got into a routine and did the same thing every day—cooking, getting dressed, reading, turning lights on, etc. He did not like the dark. The man was retired from the military and lived on a fixed income. I did not see a vehicle, so I think he was homebound.

I called my friend Steve Barnhart at Barnhart Security and told him about the strange events going on. They were curious about it, too. A technician installed an alarm system for us and then set up an elaborate motion-activated camera monitoring system that was hooked up to a computer. The system was in place for two weeks, and then the installer took it away to review it.

The technician called me and said they were surprised to find something on tape. The T.V. camera was tripped on at 1:00 AM one day by movement, and then the light in the living room came on of its own accord. You could see the switch move up on the video. This proved that something very unusual was going on. This was Barnhart's first ghost on tape, and even though no apparition showed up we all felt it was significant, and proved that there was an entity in the house. Now Barnhart offers ghost hunting surveillance services.

In 2006 my daughter, Rachel, and her fiancé' moved into this house with her three young sons. During the 1 1/2 years they lived in this house she had numerous paranormal experiences including unexplained noises, the apparition of a cat appearing on the stairs, cold spots, and other events. One evening she heard a very loud banging on the exterior walls of the lower level of the house. She thought it was someone trying to scare her as a joke, but when she looked outside she saw no one. The noise was so loud it made the house shake, and the banging shifted to different locations very quickly. After about 20 minutes of this she called the police, who did a thorough check inside and outside the house and found nothing.

Subsequently, we have had several QIG meetings at this house because it is a convenient location for all of us. We have done trance and medium work to work on missing person cases, and had good results contacting people who have passed on while working at this location. Additionally, during each of our sessions it never fails that some type of paranormal activity will occur. I suspected that there is a ley line or energy vortex at this location, and did dowsing with my L-rods. The rods, say yes, this is a vortex. So, it is no wonder we have good luck contacting spirits here.

We moved our office to this site in 2008. During the first half of 2010, my daughter, who works as my bookkeeper at night, began noticing an increase in ghostly activity at this location during the evenings. I began noticing more strange happenings at the same time during the day, as did my assistant and secretary. Some of these experiences include doors opening and closing on their own, sounds of loud footsteps and rustling papers, crashing noises, an odd rumbling/thunder-like sound where no source for it can be located, light bulbs burning out only weeks after new ones are installed, and whispering. It appears that the ghost is very active for some reason. This continues today, as of 2018.

We are planning to build a new building and move our office in order to turn this house back into a rental, however, I am concerned that any new tenants will experience the activity we have and won't stay long. In order to remove the spirit or poltergeist, I will do a cleansing of the property after we have moved out.

The Friendly Ghost at Walnut Street

This story is from Michelle M., who sent this information to me via email:

"I used to live on the 1000 block of East Walnut in Independence, Missouri. My dad sold the property around ten years ago. The house recently burned down.

My parents moved in when I was around thirteen years old. I hated the house from day one. I didn't really hate the house itself, I hated the feeling of being watched.

The house was small and cute. I was an only child. I had the entire attic to myself. It was like a separate apartment from the rest of the house and the people who'd owned it before fixed it up. It was the entire length of the house, big closet, lots of space. I should have love it.

The house had a huge yard, which I loved as well. For the most part. As an aside, and I still haven't figured this out, there was a tree near the house, and it just made me feel sad.

Anyway, after the first night, I begged to move. Mind you, I was 13. I was an only child, I didn't mind being alone. The house was close to my friend's house. I really should have loved it.

I just felt like there was a man there. I remember sometime shortly after our first night in the house, I went to my parents BAWLING and begging to move. I thought I'd heard a dog in the basement. I felt like I was relatively safe on my end of the bedroom, where my bed was, but the other end, I avoided like the plague. And I always said it was a man.

A couple of years into living there, I was having a slumber party, and my friend, Ashleigh said "there's a man down there, at the other end of your room. I just saw him."

Mind you, the room was like 10 feet wide. It wasn't like a full-grown adult man could be in my room without us seeing him, and the only adult man in the house was my dad, who wasn't going to scare the crap out of or creep on a bunch of teen girls.

When she described him, she said he was just like a shadow.

I moved out at age 17 and back in at around age 23 with my husband and our kids. The house felt weirdly better. I thought I had been an imaginative kid or, just that the new paint and wallpaper my dad had put in made a difference.

Things were fairly uneventful. I still on occasion got the feeling I wasn't alone, but I wasn't scared. It felt more like having your really cranky grandpa live with you. He didn't want you in his space, but he accepted you had to be there.

One day, I was doing dishes and my son came to me while I was washing dishes. He'd slept in the living room the night before. He said he woke up from a dead sleep and saw a man sitting on the sofa. He said the man was looking to the left, out of the window on the far side of the house. I said "I don't know babe, go outside and see what he was looking at," and went back to what I was doing. My son came in maybe five minutes later with my wedding ring in his hand. It had been missing for months, and my son said he looked in the bushes where he felt the man was looking -- it was hanging from a branch on the bush!

When we ended up moving, because my dad was selling the house, I went to the basement, which was the other creepy place in the house and said "I hope you know you don't have to stay here." It was my way of saying goodbye, and hoping "the man" would move on and be at peace.

The house caught fire about a month ago. I immediately felt bad for "the man". If he was still there, he just lost his home.

I'm also a genealogist. I've tried doing a history on this house forever, but the county records hit a brick wall in the early 80s and I knew from the property records online that it had been built in 1930. Every time I looked at the 1930 fire maps, I concluded that the house was just outside of official city limits and not listed. I couldn't figure out where in the census records to start looking.

The fire inspired me to go back and see what I could find. Finally, I cracked the case and started moving backwards in time with a city directory on Ancestry. I found the family who lived in the house originally.

Again, the house was built in 1930. It was first occupied by Charles Howe and his family. I went to look through Missouri death certs, expecting to maybe find they'd died later, but I instantly found Charles Howe's death certificate. I pulled up his information on fndagrave.com and saw that his baby son had died on the same day.

Charles Howe and his son died in a car accident the year the family moved into the house. It seemed like his kids married and moved shortly after, and his wife stayed for a few years. I don't see any other deaths linked to the house.

I could be wrong, but I now think that my ghost was Charles Howe. I feel like he was sad to see his kids move away and has hung out, confused.

I should also mention that when we first moved into the house it had just been remodeled, but in the closet in one of the bedrooms was a small toy car, which is something that you'd expect a little boy play with. I wonder if it wasn't a sign that the little boy's ghost was in the house as well.

The man who bought the house after us didn't have kids. I almost wonder if Mr. Howe didn't give up and perhaps caused the recent house fire. I should also mention that the tree in the front yard always made me feel sad, although I have no idea why."

Fairmount Park
Spirit Children

This nicely kept Fairmount park is in Sugar Creek, just north of Independence is home to more than birds and squirrels. The park sits right next to the Water Treatment plant. One evening in the fall of 2005, Janice Walker and her boyfriend, Tim Richardson, took a walk around the park, then sat on a bench next to a wooded area about 100 yards from the street at approximately 10 O'clock p.m.

After they had been sitting quietly for close to an hour, Janice and Tim heard children playing and laughing, and wondered why kids would be out so late. They hadn't noticed anyone else during their walk, and there were no cars around. The two got up to look around but couldn't see any kids. They walked further, and partly into the woods where the sounds seemed to be coming from, but couldn't find anyone around.

The couple returned in October of 2007 and went to the same bench. Once again, after about an hour, they heard the faint sounds of children laughing and playing. This time they had an idea who the children were. Two young children, Sam and Lindsay Porter, had been killed and buried in an area in the woods near the park by their father in 2004 and found in September of 2007. This unfortunate case was highly publicized, and was one that I worked on for their mother. If you go to the park at night when it is quiet, you might still hear the children playing.

The Truman Library

The Harry S. Truman Library and Museum, Independence, Missouri

500 W US Hwy 24, Independence, MO 64050
Open Monday – Saturday, 9 am – 5 pm
Sunday 12 pm – 5 pm
Closed on holidays
www.trumanlibrary.org

This presidential library is the resting place of President Harry S. Truman, the 33rd President of the United States (1945- 1953). It was the first presidential library to be created, and it was built in 1957. I was requested to speak about the haunted sites in Independence for the Truman Library in the fall of 2013. I also took the opportunity to take a tour while there and did so along with another QIG investigator Corey Pearce.

This was my second visit and tour. Prior to my visit to the site I spoke with two former employees who worked at the library for many years. One man said that the staff is well aware that strange things often occur such as people hearing their name called when no one is around, doors locking behind them when no one is in the office, sounds of footsteps walking through the halls when no one is there, and occasionally a piano playing.

One evening, a guard was standing at his post at the side door when someone tapped him on the shoulder, het when he turned to see who it was, fully expecting to see another employee, there was no one.

While I was in the building I saw the spirit of a man walking up and down the halls who used to work in engineering and who loved his job. I later found out that there was an employee who worked at the site in maintenance for many years, and that he actually died on the front steps of the building when he had a heart attack years ago. Some employees have heard strange sounds coming from the basement, where I saw the spirit.

Apparently, the main theatre, which is open to the public and used for presentations, is also haunted. An ex-staff member told me that there are often strange sounds heard coming from the stage area when no one is around. Clear footsteps walking across the stage and mechanical sounds are heard. During a speech one day, the speaker was tripped by an unseen leg when he walked across the stage.

One man I met recently used to work at the library. He was in the theatre one day when a strange multi-colored very bright four-foot-tall acorn-shaped glowing object appeared on the stage. The object was spinning. He watched this for a few minutes, then watched it fade away. He could find no explanation, and has never seen anything like it before or since.

President Truman wanted to be buried in the courtyard of the library so he could keep an eye on the place where he loved to work even after death - and perhaps he does just that!

Ron's Roadhouse Tavern

10817 E Truman Road, Independence, MO

Owners Ron and Jenelda Woolery say that this historic building may date to before 1898, and it once housed a brothel. The regulars are so used to paranormal events such as bottles that spin of their own accord, or shelves that crash to the floor are now barely noticed. But anyone who visits the second floor at Ron's Roadhouse quickly leaves. The legend of a madam named Elizabeth has been passed down from one owner to another over the years. They say that Elizabeth was killed in one of the rooms on the second floor by two of her 'Johns.' They hung her – on the site.

Even firemen checking out the building during annual inspection have left white as a ghost after visiting the second floor, so the owners keep visitors out of that area.

Elizabeth doesn't like men, it seems. Woolery had a close encounter with death when right after he purchased the bar he replaced the sign outside with the new name displayed. He suddenly felt very strange, and decided to get down and tie the ladder off. As he got halfway back up the ladder it felt like someone jerked the bottom of the ladder out from under him. Ron fell to the ground. It seems that Elizabeth would have preferred to use her name for the business rather than Ron's. Two psychics say that this is indeed the case.

Two paranormal teams, including ours, have investigated this location, and both teams say that there are several spirits lingering both upstairs and down. Two male spirits hang out at the bar and dance floor areas, and Elizabeth and a young boy seem to prefer the upper level of the building.

The Quest team plans to return to this location and do more research on the property.

Haunted Chimneys

My husband and I own a chimney contracting company, and have had a number of experiences around fireplaces and chimneys in Independence and surrounding areas. I cover events that occurred at Fort Leavenworth in the book *Haunted Greater Kansas City*, which you may find interesting. In this book I'll share more local events.

I used to write the evaluation reports for our company and saw all of the photos brought in by our technicians. I began to notice anomalies in some of the photos, which went unnoticed most of the time by the tech in the field at the time he took the photo. This occurred a number of times with different technicians. In all cases, the anomalies occurred at or near fireplaces and chimneys, but nothing was seen in other locations on the properties. Then I saw a TV show that had two episodes with spirits coming in through fireplaces. That clinched it. I wondered why this was occurring and decided to do some research on the subject, however, I could find no explanations or theories written by anyone.

One night, while meditating the answer came to me. It is DRAFT!

Ask any paranormal investigator and they will tell you that it is very common to have spirit activity around railroads, streams, and rivers, and things that cause movement. We've done experiments with fans and ductwork that were successful when contacting spirits. It is my theory that draft causes a portal to open so spirits can come and go easily, and certainly chimneys create draft- either up or down!

A stream of white fog going from a window up and out of the picture

Below: An orb was captured in back of this chimney. The homeowner related that he experienced paranormal activity around the fireplace area such as things moving and sudden big cold downdrafts when there was no wind.

Multiple orbs in different colors were captured in this photo. None of the photos before or after this one had orbs. See the color photo on my website at www.margiekay.com.

Part III
More Ghostly Things

Facts about Buying a Haunted House

How to Get Rid of Ghosts

Ghost Tours

Haunted Attractions

Events for Families

Resources

Facts about Buying a Haunted House Everyone Should Know

This information was obtained from my friend and award-winning journalist, Jason Offutt. His blog is from-the-shadows.blogspot.com. I thought my readers would find this interesting. Keep in mind that every state's laws are different.

"Stigmatized property simply means something happened on the property that could psychologically affect the buyer. That includes a natural death, murder, suicide, a previous owner had HIVAIDS, a felony was committed on the property, and in some states, ghosts.

In California, the seller has to disclose a stigma up to three years from the time it occurred. Texas has no law about it at all. But the 1991 New York appellate court tackled the supernatural in a decision (Stambovsky v. Ackley) that allowed a buyer to cancel a contract when he discovered there was a "poltergeist" in the house.

If you're buying a house in Missouri, there's a law about stigmatized property; a law that doesn't favor the buyer. It specifically states the seller isn't required to disclose a stigma –"

Why do Ghosts Haunt?

This is a question for the ages, but after doing paranormal investigations for over 30 years I've concluded that there are only a few reasons why hauntings occur. In most cases, the spirits are benevolent, and are simply watching over their loved ones, or at times, their house. Many spirits feel comfortable in the house they used to live in and are reluctant to leave. In other cases, the spirit may have unfinished business and want to communicate with someone to let them know something of importance such as where a will is located, or where a key or valuables have been hidden.

In a few instances, a person who has been murdered or killed in an accident on a property will haunt a site because they are looking for closure, justice, or even revenge. This type of spirit is especially difficult to deal with because there is so much emotion involved. The spirit may be angry, frustrated, or fearful, and when in that state of mind they are not very receptive. A psychic may be able to ask what the spirit wants, and try to deliver it, but in most cases the event happened many years ago and there is nothing that can be done today. Explaining this to the spirit and encouraging them to move on to the light and go with their spirits guides may work.

At times, spirits are not aware that they have passed over and they are just going about their daily routines. And finally, you may see a Time Imprint, which is when a "spirit" is visible, but it is simply an imprint or movie that may have been created by the materials in the walls in the house, as some theorize.

How to Get Rid of Ghosts

This is a very short outline of things to try in order to encourage spirits to leave you property. Before considering asking spirits to leave a place, think about what would happen if they were gone. Are ghosts good for business in a hotel or restaurant? If so, the owners may not want the spirits to leave. If the spirit likes being in a location, are they hurting anyone by being there? If not, maybe the best thing to do is to leave them alone. However, if the ghost is scaring people and you need to get rid of it there are a few methods that have worked well for me.

The first method is to simply ask the ghost to leave. That may sound too simple, but it usually works. You may want to talk to it and explain why by saying "You are scaring us," or "This is _____'s house and she is not comfortable having you here.", then calmly ask it to go away. If this doesn't work right away, try talking in a firmer voice without yelling or getting angry. Anger only gives more power to negative spirits

As I mentioned earlier, ghosts may not be aware that they are dead, so you may have to explain to them that they are no longer part of the physical world and that they should look around for their spirit guides and move on towards the light. I close my eyes and visualize then going into the light. This method has been very successful. Some spirits are reluctant, but if you tell them that it will be better for them this way, they eventually move on.

If you don't feel like you are communicating with the spirit, call in an experienced medium. This can usually be accomplished in one visit. If the medium finds that the spirit wants help with something, and it is something they can do, or you can do, by all means go ahead and do it. That may be the only reason the spirit is staying behind. If there is nothing you can do, have the medium communicate that you wish you could help but you can't, and it would be best for the spirit to move on.

If the spirit simply refuses, try not giving it any attention or talking about it at all. This takes energy away from the spirit. Don't show any fear, because the energy of fear empowers negative spirits. Do not use a Ouija board or have anything to do with the supernatural, which could open doors to the spirit world and allow the spirit more access to cross over to our world during the time you are trying to get it to leave.

Go about your daily routine, and when something happens that can only be attributed to a ghost try to ignore it. The ghost may give up and move on if its intentions were only to get your attention.

If it looks like the problem is more sinister, or there are multiple spirits in the house, you will probably want to engage the services of an experienced psychic, medium, ghost hunter, or even an exorcist. Find someone who has a lot of experience and check references if at all possible.

See my book "The Ghost Hunter's Field Guide" for more information about how to handle many types of entities.

Ghost Tours

If you just can't get your fill of spirits by ghost hunting on your own, here are some tours to take in Independence and the greater Kansas City area in September and October where you can experience ghosts first-hand:

Ghost Tours Fridays on the Independence Square
816-461-0065

Get tickets at the 1859 Jail.

Tours are held Fridays during October and begin at the very haunted 1859 jail. Hear about real haunted sites on the Square.

Ghosts and Gangsters Tour
816-472-4467

www.kcghostsandgangsterstour.com

Coach tour of historic Kansas City landmarks with a history of legendary paranormal activity and mafia mayhem.

Ghost Tours of haunted places in Kansas City
785-383-2925

www.ghosttourkansas.com

Visit haunted places in Kansas City including a cursed cemetery and high school. Ages 12 and over only.

Ghost Tours of Kansas
www.ghosttoursofkansas.com

Tours in Atchison, Topeka, Holton, Lawrence, Leavenworth, Manhattan, in Kansas and St. Joe. Missouri.

Kansas Ghost Works

www.kansasghostworks.homestead.com

E-mail: ksghostworks@yahoo.com

Ghost tours in ElDorado, Abiline, Shawnee, Fort Leavenworth, Fort Riley, Fort Scott, Lawrence, Topeka, and Wichita.

Haunted Atchison Trolley Tours

200 S 10th Street, Atchison Kansas

913-367-2427 or 800-234-1854

www.atchisonkansas.net

Tour the "Most Haunted Town in Kansas" on the trolley operated by the Chamber of Commerce in September and October.

Ghost Tours of Missouri

785-383-2925

www.paranormaladventureusa.com/missouri

E-mail: info@gosttourmissouri.com

Ghost tours in Excelsior Springs, Lexington, Liberty, St. Joseph, and Independence, also haunted cemeteries and battlefields.

Ghost Tours at the Ginger House Museum

Fridays in October

816-833-1602

www.margiekay.com

Email: margiekay06@yahoo.com

Ghost hunting at the site where a famous actress was born. This site is very active. Must pre-register at the website.

Haunted and Historic Spaces Tour
Downtown Lee's Summit
13 SE Third Street, Lee's Summit, Missouri
www.downtownls.org
21 and older only. Check the website for dates and times.

Wornall House
6115 Wornall Rd, Kansas City, MO
816-444-4858
www.wornallmajors.org
Take it from me, its haunted!

Fort Leavenworth Haunted Tours
Old U.S. Disciplinary Barracks
1301 N Warehouse, Fort Leavenworth, KS
www.ffam.us
One of the most haunted sites I've ever been to!

Victorian Funeral Customs
Vaile Victorian Mansion
1500 N. Liberty, Independence, MO
www.vailemansion.org
816-325-7430

Spirits from the Past
Missouri Town 1855
503-4860
Walk through the eerie village and hear spine-tingling tales over a century old. This is an annual event held in October.

Very Scary Haunted Attractions

If real haunted houses are not enough to chill you to the bone, visit any of these haunted house attractions for a scare:

KC Fear Farm
29755 W 191st Street
Gardner, Kansas
913-484-6251
www.kcfearfram.com
Not for the faint of heart. They are in the DARK and outside.

Chateau of Terror
1108 Baker Ridge Circle, Platte City, MO
Website: www.chateauofterror.com

Halloween Hallows
Jesse James Festival Grounds
704 N Jefferson St., Kearney, MO
www.halloweenhallows.com

Macabre Cinema
1222 W 12th St., Kansas City, MO
816-842-0320
More than 30 scenes throughout this four-floor haunted 1930's style movie theatre madness.

The Chambers of Edgar Allen Poe

1100 Santa Fe, Kansas City, MO

816-842-0320

www.chambersofpoe.com

Based on the tales of Edgar Allen Poe with scenes of The Raven, Rue Morgue, The House of Usher, The Black Cat, and more. Experience the feelings of suffocating, claustrophobia, or being buried alive. Very scary!

The Beast

1401 W 13th St., Kansas City, MO

816-842-0320

www.kcbeast.com

One of the oldest, largest, and scariest haunted houses in the K.C. area. I'd suggest adults only.

The Edge of Hell

1300 W 12th St. Kansas City, MO

816-842-0620

www.edgeofhell.com

KC's oldest haunted house with a quarter-mile walk through and a slide at the end. Too scary for children under 12.

The Middle of Nowhere

The Castle of Dr. Cranken-Hertz

506 E Forest St., Harrisonville, MO

www.middleofnowherehaunt.com

Worlds of Fun Halloween Haunt

Weekends in September and October

The Living Dead Haunt

1021 NE Colbern Rd
Lee's Summit, MO 64086
816-600-6300

Mount Washington Manor Haunted House

9515 E Independence, Ave., Independence, MO
816-866-3379
www.mtwashingtonmanor.com
Open Friday and Saturdays in October, and the last weekend of
September. This haunted house is in the lower level of a 95-year old
Masonic temple. Family friendly but scary. A community fundraiser.

Events for Families

To get the kids trained for more frightening stuff when they get older, start by taking them to these no-scary fall events:

Enchanted Forest
George Owens Nature Park
1601 S. Speck Road, Independence, MO
816-325-7370
Non-scary Halloween event for families with storybook and nursery rhyme characters, wagon hayride and lighted display.

Independence Halloween Parade
Annual Halloween Parade in Independence, watch the papers for details.

BOO-Springs Harvest Fest
Downtown Blue Springs on Main Street
228-0137

Nox Noctis
1859 Mallard Dr., Liberty, MO
www.cutcliffe.com/Darkness.htm
A free haunted house in a garage in Liberty for kids.

World's of Fun Halloween Haunt
4545 Worlds of Fun Ave., Kansas City, MO
www.haunt.worldsoffun.com
For kids and teens.

Spirits from the Past

Missouri Town 1855

503-4860

Walk through the eerie village and hear spine-tingling tales over a century old. This is an annual event held in October.

KC Pumpkin Patch

29755 W 191st St., Gardner, Kansas

913-484-6251

www.kcpmmpkinpatch.com

Pumpkin Patch Fall festival, corn maze, tractor ride, and more than 30 fun family activities and festival foods.

Corn Maze

Unique corn maze at the KC Pumpkin Patch (see above)

Weston Red Barn Farm

16300 Wilkerson Rd. Weston, Missouri

816-386-5437

Fall festival, hayrides, pick pumpkins and apples.

Faulkner's Ranch & Pumpkin Farm

10600 Raytown Rd., Kansas City, MO

816-761-5055

www.faulknersranch.com

Fall festival, hayride, field maze, pumpkins, fall produce.

Powell Gardens

Events for families with glowing gardens, food, music, and activities throughout the fall.

Louisburg Cider Mill

Lousiburg, Kansas

Corn maze and pumpkin patch throughout the season and an annual burning scarecrow, music, hayrides, and tour through the maze in the dark.

Resources for Investigation Purposes

Midwest Genealogy Center
3440 S Lee's Summit Road
Independence, MO 64055
816-252-7228
Hours Mon-Thurs 9-9, Fri 9-6, Sat 9-5, Sun 1-5 except for holidays

Jackson County Historical Society
Jackson County Truman Courthouse on the Square
816-461-1897
www.jchs.org
By appointment only Tuesday through Friday.

Mid-Continent Public Library
15616 E US 24 Hwy
Independence, MO
816-455-5030
www.mymcpl.org
Newspaper archives, genealogy research.

Ancestry.com
Newspaper archives, war records, death index.

Jackson County Records
www.jacksongov.org

Bibliography

Kay, Margie. 2013 *Gateway to the Dead: The Ghost Hunter's Field Guide.* *Nocturna Press.*

Christine, Maria. 2010 *Phantom Encounters: Chillingly True Ghost Stories.* Create Space.

Gilbert, Joan. *Missouri Ghosts*

Offutt, Jason. 2007 *Haunted Missouri: A Ghostly Guide to Missouri's Most Spirited Spots.* Truman State University Press.

Offutt, Jason. 2010 *"What Lurks Beyond: The Paranormal in Your Backyard.* Truman State University Press

The Kansas City Star

Websites

Paranormal Research Society, paranormalinvestigators.com

Larry Moore, KMBC's Ghost Stories videos on You Tube, www.kmbc.com.

Missouri Paranormal, www.missouriparanormal.net

Millers Paranormal Research, www.millersparanormalresearch.com

Paranormal Missouri, www.paranormalmissouri.com

Wikipedia, www.wikipedia.com

The Shadow Lands: www.theshadowlands.net/places/missouri.htm

Legends of America: www.legendsofamerical.com/mo-independencehauntings.html

Haunted Houses.com: www.hauntedhouses.com

The FBI website: www.fbi.gov

If you know of any haunted sites that should be investigated please contact Margie Kay at www.margiekay.com or margiekay06@yahoo.com

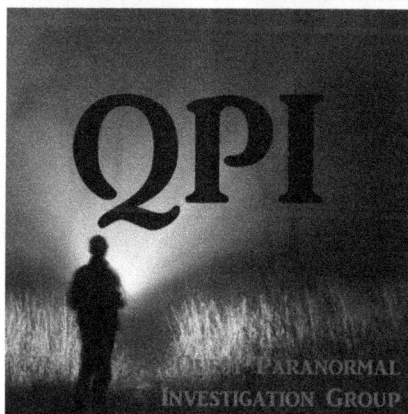

QUEST

Investigation Group

- Haunted house investigations
- Haunted house certifications
- Lectures
- Free assistance with unsolved crimes
- Free assistance with missing persons

Contact: MargieKay06@yahoo.com

Books and Films by Un-X News Media

Haunted Independence by Margie Kay (second edition) 2019
The Kansas City UFO Flaps by Margie Kay 2018
Gateway to the Dead: A Ghost Hunter's Field Guide by Margie Kay 2013
The UFO Hunter's Field Guide by Margie Kay 2019
Color Therapy Wall Chart by Margie Kay 1999
Family Secrets by Jean Walker 2017
All available at Amazon.com

Coming soon:
Welcome to Earth: book and documentary film by Margie Kay
Real People - Real UFOs by Margie Kay
ExtraOrdninary by Margie Kay
Paranormal Hot Spot: Missouri by Margie Kay
www.margiekay.com
www.unxmedia.com

Books by Margie's Daughter, Maria Christine
Chillingly True Ghost Stories 2007
Parallel: New Beginnings and Old Ghosts (fiction) 2010
The Seventh Dimension (fiction) 2011
Phantom Encounters 2018
www.mariachristineonline.com

About the Author

Margie Kay is a native of the Kansas City area and lives in Independence, Missouri with her husband, Gene. Together they own a chimney contracting business, a real estate investment company, and a publishing company. Kay holds several technical certifications. Margie is a nationally known speaker, and has written numerous articles for magazines and newspapers. She is the author of twelve books and several documentary films. Margie was host of QUEST radio show on KCXL radio for five years and Un-X News on KGRAradio.com for two years.

Kay has been consulted by other authors and researchers, and her comments appear in several books and films. She and her daughter appear in the TV pilot "Strange," which is currently on YouTube. Margie's interest in the paranormal began at a very early age, and she showed psychic abilities before the age of three. Kay's natural abilities to see through objects and people, predict events, and do remote real-time or past viewing have astounded many.

Kay's talents have led to solving over 50 missing person and homicide cases. Law enforcement and private investigators from all over the U.S. call on her to help with difficult cases. She is the founder and director of QUEST Investigation Group, which is a non-profit association organized to help find missing persons and do paranormal investigations.

www.ingramcontent.com/pod-product-compliance
Lightning Source LLC
Chambersburg PA
CBHW022057210326
41519CB00054B/546